KRIGER'S TEXTBOOK OF PHARMACOLOGY

ISBN: 978-1-105-64212-8

Published in Canada by Altaspera Publishing & Literary Agency Inc.

ISBN 978-1-105-64212-8

Contents

GENERAL PRINCIPLES OF PHARMACOLOGY

DRUG INPUT AND DISPOSITION

Drugs are compounds almost always foreign to the body. As such, they are not continually being formed and eliminated as are endogenous substances. The processes of inputting, distributing, and eliminating drugs are therefore of paramount importance in determining the onset, duration, and intensity of effect.

DRUG ABSORPTION

The process of drug movement from the site of administration toward the systemic circulation.

Drug product: The actual dosage form of a drug, consisting of the drug itself plus other ingredients formulated into a usable medicine; eg, as a tablet, capsule, or solution. Drug products are formulated for administration by a variety of routes, including oral, buccal, sublingual, rectal, parenteral, topical, and inhalational. The physicochemical properties of drugs, their formulations, and the routes of administration are important in absorption. A prerequisite to absorption of any drug is that it be able to enter into solution. The solid drug product (eg, tablet) must undergo disintegration and deaggregation, and the active ingredients must undergo dissolution before the drug can be absorbed.

Except when given IV (intravenously), a drug must traverse several semipermeable cell membranes before reaching the general circulation. These membranes act as biologic barriers that selectively inhibit the passage of drug molecules. Cell membranes are composed primarily of a bimolecular lipid matrix, containing mostly cholesterol and phospholipids, in which are embedded globular protein macromolecules of random size and composition. The membrane proteins may be involved in transport processes and may also function as receptors for cellular regulatory mechanisms. Membrane lipid provides stability to the membrane and determines its permeability characteristics.

Processes

The processes by which drugs move across a biologic barrier include passive diffusion, facilitated diffusion, active transport, and pinocytosis.

Passive diffusion: Transport across a cell membrane in which the driving force for movement is the concentration gradient of the solute. Most drug molecules are transported across a membrane by simple diffusion from a high concentration area (GI fluids) to a low concentration area (blood) without expenditure of energy. The net rate of diffusion is directly proportional to this net gradient and depends upon lipid solubility, degree of ionization, molecular size, and the area of the absorptive surface. Since the drug is rapidly removed by the systemic circulation and distributed into a large volume, the concentration of drug in blood is initially low compared with that at the site of administration. The resulting large concentration gradient serves as the driving force for absorption. However, since the cell membrane is lipoidal in nature, drugs that are lipid soluble diffuse more rapidly than drugs that are relatively lipid insoluble. Furthermore, small molecules tend to penetrate membranes more rapidly than do large ones.

Most drugs exist as weak organic acids or bases in both nonionized and ionized forms in an aqueous environment. The nonionized fraction is usually lipid soluble and diffuses readily across cell membranes. The ionized form cannot penetrate the cell membrane easily because of its low lipid solubility.

Facilitated diffusion: For certain molecules (glucose), the rates of penetration are greater than expected from their low lipid solubility and the concentration gradients present. It is postulated that a "carrier component" combines reversibly with the substrate molecule at the cell membrane exterior and that the carrier-substrate complex diffuses rapidly across the membrane with release of the substrate at the interior surface. This carrier-mediated diffusion process is characterized by selectivity and saturability. The carrier mechanism accepts for transport only those substrates having a relatively specific molecular configuration, and the process is limited by the availability of carrier. No expenditure of energy is required by this process; substrate is not transported against a concentration gradient.

Active transport: In addition to selectivity and saturability, active transport requires energy expenditure by the cell, and substrates may accumulate intracellularly against a concentration gradient. Active transport processes appear to be limited to agents with structural similarities to normal body constituents. These agents are usually absorbed from specific sites in the small intestine. Active transport processes have been identified for various ions, vitamins, sugars, and amino acids.

Pinocytosis refers to the engulfing of particles or fluid by a cell. The cell membrane invaginates (forms and opening), encloses the particle or solute, and then fuses again, forming a vesicle that later buds off within the interior of the cell. This mechanism also requires the expenditure of energy. Pinocytosis probably plays a minor role in drug transport.

Oral Administration

Because the oral route of administration is the most common, absorption usually refers to the transport of drugs across the membranes of the epithelial cells within the GI tract. Absorption after oral administration is confounded by differences down the alimentary canal in the luminal pH; surface area per luminal volume; perfusion of the tissue, bile, and mucus flow; and the epithelial membranes. The faster absorption of acids in the intestine compared with the stomach appears to contradict the hypothesis that the nonionized form of a drug more readily crosses membranes.

Gastric emptying and intestinal transit time: Because the absorption of virtually all compounds is faster from the small intestine than from the stomach, the rate of gastric emptying is a controlling step. Food, especially fatty foods, slows gastric emptying, which explains why some drugs are recommended to be taken on an empty stomach when a rapid onset of action is desired. The extent of absorption may be enhanced by food if the drug is poorly soluble (e.g., griseofulvin) or reduced if degraded in the stomach (e.g., penicillin G). Drugs that affect gastric emptying (parasympatholytic agents) also affect the rate of absorption of other drugs.

Parenteral Administration

Direct placement of a drug into the bloodstream (usually IV) ensures complete delivery of the dose to the general circulation. However, administration by a route that requires drug transfer through one or more biologic membranes to reach the bloodstream precludes a guarantee that all of the drug will eventually be absorbed. Intra-muscular (IM) or Sub-cutaneous (S.C.). injection of drugs bypasses the skin barrier, but the drug must penetrate the capillary walls. Because the capillaries tend to be highly porous, the perfusion (blood flow/gram of tissue) is a major factor in the rate of absorption. Thus, the injection site can markedly influence a drug's absorption rate.

Controlled-Release Dosage Forms

Controlled-release dosage forms are designed to reduce the frequency of dosing and to maintain more uniform plasma drug concentrations, thus providing a more uniform pharmacologic effect. Reduction of the absorption rate can be achieved in various ways: by coating the drug particles with wax or related water-insoluble material, by embedding the drug in a matrix from which it is released slowly during transit through the GI tract, or by complexing the drug with ion-exchange resins.

Topical controlled-release dosage forms have been designed to provide drug release for extended periods; eg, clonidine diffusion through a membrane provides controlled drug delivery over a period of 1 wk, and nitroglycerin-impregnated polymer bonded to an adhesive bandage provides controlled drug delivery over a period of 24 h. Drugs for transdermal delivery must have suitable skin penetration characteristics and high potency.

BIOAVAILABILITY

The rate at which and the extent to which the active moiety (drug or metabolite) enters the general circulation, thereby gaining access to the site of action.

Although a drug may be absorbed completely, its rate of absorption may also be important. It may be too slow to attain a therapeutic blood level in an acceptable period of time or too rapid, resulting in toxicity from high drug levels just after each dose.

Causes of Low Values of Bioavailability

When a drug rapidly dissolves from a drug product and readily passes across membranes, absorption from most sites of administration tends to be complete. This is not always the case for drugs given orally. Before reaching the vena cava, a drug must move down the alimentary canal and pass through the gut wall and liver, which are common sites of drug metabolism ,thus, the drug may be metabolized before it can be measured in the general circulation. This cause of a decrease in drug input is called the first-pass effect. A large number of drugs show low bioavailabilities owing to extensive first-pass metabolism. In many instances, the extraction is so complete that the bioavailability is virtually zero (isoproterenol, norepinephrine, phenacetin, and testosterone).

The 2 other most frequent causes of low bioavailability are an insufficient time in the GI tract and the presence of competing reactions. Ingested drug is exposed to the entire GI tract for no more than 1 to 2 days and to the small intestine for only 2 to 4 h, unless gastric emptying is considerably delayed. If the drug does not dissolve readily or if the drug is incapable of penetrating the epithelial membrane (highly ionized and polar), there may be insufficient time at the absorption site. Not only is the bioavailability low in this case, but it tends to be highly variable. In addition, individual variations in age, sex, activity, genetic phenotype, stress, disease (achlorhydria, malabsorption syndromes), and previous GI surgery can alter and further increase variability in drug bioavailability.

Reactions that compete with absorption can reduce bioavailability - include complex formation; hydrolysis by gastric pH or digestive enzymes ;conjugation in gut wall ; adsorption to other drugs and metabolism by luminal microflora.

DRUG DISTRIBUTION

After a drug enters the general circulation, it distributes throughout the body's tissues. Distribution is generally uneven because of differences in binding in tissues, regional variations in pH, and differences in the permeability of cellular membranes.

Binding Components

The extent of the distribution of drugs into tissues depends on binding to plasma proteins and tissue components.

Plasma protein binding: Drugs are transported in the bloodstream partly in solution (as free drug) and partly bound to various blood components (plasma proteins and blood cells). Many plasma proteins can interact with drugs. Albumin, alpha1-acid glycoprotein, and lipoproteins are the most important ones. Acidic drugs are generally bound more extensively to albumin, while basic drugs often are more extensively bound to either one or both of the latter 2 proteins.

Because only the unbound form is available for passive diffusion to the extravascular or tissue sites where pharmacologic effects occur, plasma protein binding influences the distribution and apparent relationship between pharmacologic activity and plasma (total) drug concentration.

The fraction unbound (Fu) is often more useful than the fraction bound.

Tissue Binding

The substances to which drugs bind in tissue are highly varied. Often these substances are not proteins. Furthermore, they may be very specific, as is the case for the binding of chloroquine to nucleic acids. Tissue binding usually involves an association of drug with a macromolecule within an aqueous environment. Another kind of association that results in apparent tissue binding is partitioning of drug into body fat.

Drug reservoir: Accumulation of drugs in tissues or body compartments can prolong drug action because the tissues serve as depots.

Some drugs accumulate in cells in higher concentrations than those in Extra Cellular Fluid (ECF). Such accumulation most commonly involves binding of drugs with protein, phospholipids, or nucleic acids.

Passage of drugs into the Central Nervous System (CNS) takes place in the capillary circulation and the Cerebro Spinal Fluid (CSF). Although the brain receives a large proportion of the cardiac output (about 1/6), distribution of drugs to brain tissue is restricted. While some lipid-soluble drugs (e.g., thiopental) do enter and exert their pharmacologic effects rapidly, many drugs, particularly the more water-soluble agents, enter the brain slowly. Another important barrier to water-soluble substances is the close approximation of the glial connective tissue cells (astrocytes) to the basement membrane of the capillary endothelium. The capillary endothelium and the astrocytic sheath together are referred to as the blood-brain barrier.

DRUG ELIMINATION

The sum of the processes of drug loss from the body. Removal of drugs from the body occurs by metabolism and excretion.

METABOLISM

The process of chemical alteration of drugs in the body. The liver is the principal, but not the sole, site of drug metabolism. Some metabolites are pharmacologically active. When the substance administered is inactive and an active metabolite is produced, the administered compound is called a pro-drug.

Changes with Age

Neonates have partially developed liver microsomal enzyme systems and, consequently, have difficulty with the metabolism of many drugs. Elderly patients often show a reduced ability to metabolize drugs. The reduction varies depending on the drug and is not as severe as that in neonates.

Individual Variation

Variability among individuals makes it difficult to predict the clinical response to a given dose of a drug. Some patients may metabolize a drug so rapidly that therapeutically effective blood and tissue levels are not achieved; in others, metabolism may be so slow that toxic effects result with usual doses. Concurrent disease states, particularly chronic liver disease, drug interactions, especially those involving induction or inhibition of metabolism, and other factors also contribute.

DRUG TOXICITY

PRECLINICAL AND CLINICAL EVALUATION OF TOXICITY

Before a drug is approved for general clinical use by the FDA, preclinical and clinical data showing substantial evidence of safety and efficacy are required by law. Drug studies proceed through various phases, as follows.

Preclinical Investigation (Animal Studies)

Animal studies used to determine or define the safety of a drug include studies of acute, subchronic, and chronic toxicity in several animal species.

The initial acute toxicity studies are to determine the median lethal dose (LD50), the toxic symptoms developed by the animals, and the time that they appear. At least 3 species of animals, one not a rodent, are usually used, and acute toxicity is usually determined by more than one route of administration.

Subchronic toxicity studies are conducted in at least 2 animal species and usually consist of daily administration of the test drug for up to 90 days. In each species, at least 3 dose levels are used, varying from the expected therapeutic doses to levels high enough to produce toxicity.

Chronic toxicity studies are carried out in at least 2 species, one of which is not a rodent. These studies usually last for up to the lifetime of the animal, but their length will depend on the intended duration of administration of the drug to humans.

Clinical Investigation (Human Studies)

Some adverse effects of drugs cannot be discerned in animals; e.g., dizziness, nausea, headaches, ringing in the ears, heartburn, and depression. It has been estimated that $>= 50\%$ of undesirable drug effects seen most frequently can be ascertained only during human trials.

Phase 1 represents the first administration of a new drug to man. A small number of closely monitored subjects, mainly healthy volunteers, are usually involved. Initially, each receives a single dose of the drug to determine a safe dose range and assess pharmacokinetic data. The primary objective of this necessarily cautious phase of the investigation is to determine a safe and tolerated dosage in humans; however, observation of toxicity, if it occurs, and of absorption, metabolism, and excretion may also be made during Phase 1.

Phase 2 begins after satisfactory preliminary evidence regarding safety has been obtained. It involves the supervised administration of the drug to patients for treatment of, or prophylaxis against, the disease or symptoms for which the drug is intended. These studies usually are conducted in randomized clinical trials comparing the new drug with the prototype drug, if any, for a particular indication. Often this is the first opportunity to observe the effect of long-term administration of the drug to humans.

Phase 3 begins after the initial phases have provided reasonable evidence of safety and efficacy. It consists of more widespread clinical trials that may move from the realm of clinical investigators to practicing physicians. Phase 3 extends up to the time the drug is released for general use.

Phase 4 is the study of the actual use of the drug in medical practice and, though often not recognized as a phase of clinical investigation, is a most important one from a clinical standpoint

ADVERSE DRUG REACTIONS (ADRs)

ADRs are usually classified as mild (no antidote, therapy, or prolongation of hospitalization necessary); moderate (requires a change in drug therapy, although not necessarily cessation of the drug, and may prolong hospitalization or require special treatment); severe (potentially life-threatening, requires discontinuation of the drug and specific treatment of the adverse reaction); and lethal (directly or indirectly contributes to the death of the patient).

Dose-Related, Predictable Drug Reactions

Side effects are predictable pharmacologic effects that occur within therapeutic dose ranges and are undesired in the given therapeutic situation. Side effects may be useful under certain circumstances.

Overdosage toxicity is the predictable toxic effect that occurs with dosages in excess of the therapeutic range for a particular patient. It overlaps with side-effect toxicity to some extent, especially in drugs with a small therapeutic index. The severity of the reaction is usually dose-related.

Non-Dose-Related, Unpredictable Effects

Drug allergy: Allergic reactions depend on altered reactivity of the patient as a result of prior contact with a drug that functions as an antigen or allergen. They are not dose-related; the symptoms and signs that develop are determined by antigen-antibody interactions and are largely independent of the pharmacologic properties of the drug. Allergic reactions are not completely unpredictable; a careful clinical history may suggest at risk.

Idiosyncrasy is an imprecise term that has been used as a classification for unexpected and peculiar adverse reactions occurring in a small percentage of individuals exposed to a drug. Idiosyncratic reactions are not related to a drug's known pharmacologic effects and are not obviously allergic in nature. Idiosyncrasy has been defined by some as a genetically determined abnormal reactivity to a drug.

DRUGS IN PREGNANCY

Drugs given during pregnancy can affect the fetus by (1) acting directly on the embryo to produce a lethal, toxic, or teratogenic effect; (2) altering placental function (constricting vessels), affecting gas and nutrition exchange between fetus and mother; (3) changing the myometrial activity (producing severe uterine hypertonia resulting in fetal anoxic injury); or (4) altering the biochemical dynamics of the mother, indirectly affecting the fetus.

There is 5 categories of safety for use in pregnancy. In category A, controlled human studies have demonstrated no fetal risks (these are the safest drugs). In category B, animal studies indicate no risk to the fetus and no controlled human studies have been done, or animal studies show a risk to the fetus but well-controlled human studies do not. In category C, no adequate studies, either animal or human, have been done, or adverse fetal effects have been shown in animals but no human data are available. In category D, positive evidence of human fetal risk exists, but benefits in certain situations (e.g., life-threatening situations or serious diseases for which safer drugs cannot be used or are ineffective) may outweigh the risks. In category X, proven fetal risks exist that outweigh any possible benefit. These labeling definitions are universally accepted and are often helpful in directing the risk-benefit decision making encountered when prescribing drugs during pregnancy.

Teratogens and fetal inrtoxicants

Antineoplastic agents, Isotretinoin, androgenic hormones and synthetic progestins, thyroid drugs: radioactive iodine, triiodothyronine, propylthiouracil, methimazole; oral hypoglycemics, narcotics, sedatives, alcohol, tranquilizers and antidepressants, tetracyclines, long-acting sulfonamides, anticoagulants (Coumarins), caffeine, cigarette smoking.

PRINCIPLES OF PHARMACODYNAMICS, TRANSDUCTION AND NEUROTRANSMISSION

A nerve cell or neuron has 2 major distinguishing functions - propagation of the action potential along the axon, and transmission of the signal from one nerve to another nerve or cell structure to elicit a response (eg, nerve impulse, muscle contraction). While impulses conducted along an axon are caused by the movement of Na and K ions across the membrane and are electrical in nature, the transmission of impulses from one neuron to another neuron or to a non-neuronal cell is chemical and depends upon the action of certain neurotransmitters (NTs) on specific receptors.

Transmission via NTs is a highly complex and sensitive process. Synaptic relationships in the periphery involve neuron-neuron or neuron-effector interactions. Neurotransmission can be increased or decreased to accommodate any physiologic situation. Many neurologic and psychiatric diseases are caused by a pathologic over- or under activity of neuronal transmission. Many drugs can modify neurotransmission to cause adverse effects (e.g., hallucinogens) or to correct pathologic conditions (e.g., antipsychotic drugs).

Basic Principles of Neurotransmission and transduction

Neurotransmission involves (1) synthesis and storage of the NT in the prejunctional nerve structure; (2) release of the NT from the nerve terminal; (3) interaction of the NT with a specific postjunctional structure (receptor); (4) rapid termination of the NT-receptor interaction; and (5) destruction of NT or re-uptake into the terminal.

Regulation of NT amounts varies among neurons but is achieved through increased or decreased precursor uptake or activity of NT synthesizing or destroying enzymes. Also, stimulation or blockade of postsynaptic receptors can decrease or increase presynaptic NT synthesis.

Major Neurotransmitters

A neurotransmitter (NT) is defined as a chemical that is selectively released from a nerve terminal by an action potential, interacts with a specific receptor on an adjacent structure, and elicits a specific physiologic response.

Most NTs derive from amino acids (or related compounds such as choline). Certain neurons synthesize only one, neuron-specific NT; others have been shown to synthesize 2 or more NTs. Some neurons modify amino acids to form the "amine" transmitters (e.g., norepinephrine, serotonin, acetylcholine); others combine amino acids to form "peptide" transmitters (e.g., endorphins, enkephalins); and still other neurons use amino acids unchanged or synthesized as transmitters. A few NTs are not related to amino acids.

Acetylcholine (Ach), the major NT of the motoneurons, autonomic preganglionic fibers, postganglionic cholinergic (parasympathetic) fibers, and many neurons in the CNS (basal ganglia, motor cortex), is synthesized from choline and mitochondrially derived acetyl-coenzyme A by the enzyme choline acetyltransferase (CAT). Upon release, Ach stimulates cholinergic receptors of adjacent structures. This interaction is rapidly terminated by hydrolysis of Ach to choline and acetate by the enzyme acetylcholinesterase (ACE) found adjacent to the receptors. Ach levels are regulated by the activity of CAT and by choline uptake.

Dopamine (DA) is the NT of some peripheral nerve fibers and of many central neurons (e.g., substantia nigra, midbrain, hypothalamus). The amino acid tyrosine is taken up by dopaminergic neurons, converted by the enzyme tyrosine hydroxylase to 3,4-dihydroxyphenylalanine (dopa), decarboxylated by the enzyme aromatic l -amino acid decarboxylase to DA, and stored in vesicles. After release, DA interacts with dopaminergic receptors and is then "pumped" back by active processes (re-uptake) into the prejunctional neurons. DA levels are held constant by changes in tyrosine hydroxylase activity and the enzyme monoamine oxidase (MAO), which is localized in nerve terminals and metabolizes dopamine. DA is metabolized to several metabolites, including specifically homovanillic acid.

Diseases that affect the function of signal transmission can have serious consequences. Parkinson's disease has a deficiency of the neurotransmitter dopamine. Progressive death of brain cells increases this deficit, causing tremors, rigidity and unstable posture. L-dopa is a chemical related to dopamine that eases some of the symptoms (by acting as a substitute neurotransmitter) but cannot reverse the progression of the disease.

Norepinephrine (NE) is the NT of most postganglionic sympathetic fibers and many central neurons (e.g., locus ceruleus, hypothalamus). NE synthesis, like that of DA, also starts with the precursor tyrosine but continues as DA is hydroxylated by dopamine-beta-hydroxylase to form NE, which is stored in vesicles. Upon release, NE interacts with adrenergic receptors. This action is terminated largely by the re-uptake of NE back into the prejunctional neurons. Tyrosine hydroxylase and MAO regulate intraneuronal NE levels. Metabolism of NE occurs via MAO and catechol-O-methyltransferase to inactive metabolites (e.g., normetanephrine, 3-methoxy-4-hydroxyphenylethylene glycol, 3-methoxy-4-hydroxymandelic acid).

Serotonin (5-HT) is the NT of many central neurons. Its synthesis begins with the uptake of tryptophan into serotonergic neurons. Tryptophan is hydroxylated by the enzyme tryptophan hydroxylase to 5-hydroxytryptophan, and then decarboxylated to serotonin (5-hydroxytryptamine) by the enzyme aromatic 1-amino acid decarboxylase. Levels of 5-HT are controlled by the uptake of tryptophan and intraneuronal MAO. Metabolism occurs mainly via MAO to 5-hydroxyindoleacetic acid.

Gamma-Aminobutyric acid (GABA) causes mostly inhibitory responses in the CNS and is found in many areas (e.g., basal ganglia, cerebellum). GABA is derived from glutamic acid, which is decarboxylated by glutamic acid decarboxylase. After interaction with its receptors, GABA is actively "pumped" back into the neuronal terminals. It is metabolized by a GABA-transaminase.

The bacterium Clostridium tetani produces a toxin that prevents the release of GABA. GABA is important in control of skeletal muscles. Without this control chemical, regulation of muscle contraction is lost; it can be fatal when it effects the muscles used in breathing.

Clostridium botulinum produces a toxin found in improperly canned foods. This toxin causes the progressive relaxation of muscles, and can be fatal. A wide range of drugs also operate in the synapses: cocaine, LSD, caffeine, and insecticides.

β-Endorphin (β-End) is a polypeptide and is the transmitter of many central neurons (e.g., hypothalamus, amygdala, thalamus, locus ceruleus). After release and interaction with peptidergic (opioid) receptors, it is hydrolyzed by peptidases into smaller, inactive peptides and amino acids.

Methionine-enkephalin and leucine-enkephalin are small peptides present in many central neurons (e.g., globus pallidus, thalamus, caudate, central gray). After release and interaction with peptidergic (opioid) receptors, the enkephalins are hydrolyzed by other peptidases into smaller, inactive peptides and amino acids.

Dynorphins are a group of 7 peptides with similar amino acid sequences and are present in the same areas as are the enkephalins. These peptides are derived from prodynorphin and are hydrolyzed after receptor activation.

Substance P is a peptide and the transmitter of many central neurons (e.g., dorsal root ganglia, basal ganglia, hypothalamus). Its synthesis and fate are similar to those of the other peptide NTs.

Glycine, glutamate, and aspartate are NTs used directly by certain neurons without change (although glycine might also be synthesized from serine). Aspartate is mainly present in the cortex, glutamate in the cerebellum and spinal cord, and glycine in the interneurons of the spinal cord. Glutamic and aspartic acid cause excitatory responses, while glycine is inhibitory.

Other neurotransmitters, whose roles in neurotransmission have not been as firmly established, include epinephrine, histamine, vasopressin, vasoactive intestinal peptide, carnosine, bradykinin, cholecystokinin, bombesin, somatostatin, corticotropin-releasing factor, neurotensin, and others.

In addition to these amino acid-related NTs, some NTs are different, e.g., adenosine.

Major Receptors

A receptor is a binding or recognition site with a specific molecular configuration on a cell surface or within the cell structure, which causes a physiologic response upon stimulation by an NT or other chemical, such as a drug or toxin. Some receptors cause inhibitory (e.g., relaxation of a muscle) or excitatory (e.g., initiation of nerve impulse or contraction of a muscle) responses. Usually, movement of $Na+$ depolarizes and is stimulatory, whereas movement of Cl^- hyperpolarizes and is inhibitory.

Ion channel receptors can be classified into receptors that are channel (e.g., nicotine, GABA, glycine, and glutamate receptors) or second messenger receptors that are activated by a second messenger to affect the channel (e.g., adrenergic, muscarinic, serotonergic, and dopaminergic receptors).

Receptors are continuously synthesized in the cell body, transported to the respective sites, stored in the membrane, and after use degraded. Their half-lives range from days to weeks. The number of receptors and their affinity for specific NT molecules is not constant. Receptors that are continuously stimulated by NTs or drugs (agonists) become hyposensitive ("down-regulation"), whereas receptors that are not stimulated by their NT or are blocked by drugs (antagonists) become hypersensitive ("up-regulation"). Up-regulation or down-regulation of receptors plays a major role in the development of tolerance and physical dependence (pharmacodynamic tolerance or dependence). Withdrawal is usually a rebound phenomenon due to an altered receptor affinity and/or density.

Most NTs interact primarily with the postsynaptic receptor (R-1) to produce a physiologic response in the adjacent structure. However, receptors are also located on presynaptic neurons and control the release of a specific NT. These receptors can be divided into different classes. The autoreceptors (R-2) respond only to released NT and are of 2 types: (R-2,+) that increase the release of NT, and

(R-2,) that inhibit the release of NT. Presynaptic receptors (R-3) can increase or inhibit the release of an NT. In addition, receptors for neuromodulators (R-4) or substances that are not released from nerve terminals (e.g., steroids, prostaglandins) can modulate the release of the NT.

Cholinergic receptors can be divided into nicotinic N1 (adrenal medulla, autonomic ganglia) and N2 (skeletal muscle) receptors as well as muscarinic M1 (autonomic system, striatum, cortex, hippocampus) and M2 (autonomic system, heart, intestinal smooth muscle, hindbrain, cerebellum) receptors

Adrenergic receptors can be divided into alpha1 (postsynaptic in the sympathetic system) and alpha2 (presynaptic in the sympathetic system and postsynaptic in the brain) receptors, as well as β-1 (heart) and β-2 (other sympathetically innervated structures) receptors.

Dopaminergic receptors can be classified as D1, D2, and D3 receptors. D1 receptors activate adenylate cyclase via stimulatory G-proteins, whereas D2 receptors inhibit this enzyme via inhibitory G-proteins. D1 receptors are more frequent than D2 receptors (4:1), but both receptors are formed in the same brain areas (e.g., limbic region, basal ganglia). The D3 receptor does not seem to affect adenylate cyclase and is more localized in the limbic areas. In addition, isoforms of the individual receptors have been detected.

GABA receptors can be divided into GABAA receptors activating chloride channels, or GABAB receptors potentiating adenosine 3,5-cyclic phosphate (cAMP) formation. This site can be influenced by benzodiazepines (eg, benzodiazepine binding increases GABA binding), barbiturates, picrotoxin, or muscimol.

Serotonin (5-HT) receptors can be divided into 5-HT1, 5-HT2, and 5-HT3 receptors.

Glutamate receptors (excitatory) can be subclassified as N-methyl-D-aspartate (affecting the flow of Na+, K+, Ca+ +), quisqualate (Na+, K+), and kainate (Na+, K+) receptors.

Endorphin-enkephalin or opioid receptors can be divided into μ1 and μ2 (sensorimotor integration, analgesia), delta (motor integration, cognitive function), and kappa 1 and kappa 2 (water balance regulation, analgesia, food intake). All receptors are inhibitory in nature, are often located presynaptically, and seem to be coupled to G-proteins.

Second Messenger Systems

Second messenger systems are complexes of regulatory (eg, G-proteins) and catalytic (eg, adenylate cyclase, phospholipase C) proteins, which are activated by NTs (first messengers) to form specific chemicals or second messengers (eg, cAMP, inositol triphosphate or IP3, diacylglycerol or DAG).

Adenylate cyclase-cAMP is the best known second messenger system. Here, the first messenger (NT, hormone) binds to the receptor activating a stimulatory G-protein (Gs) by displacing guanosine diphosphate (GDP) with guanosine triphosphate (GTP). G-proteins consist of alpha, beta, and gamma subunits; the alpha unit binds the guanine nucleotide and provides specificity for receptors. The activated protein amplifies the signal of the first messenger and activates adenylate cyclase. This enzyme converts adenosine triphosphate (ATP) to cAMP, which activates specific phosphorylating enzymes or protein kinases to produce the physiologic response. The action of cAMP is terminated by the enzyme phosphodiesterase. In addition to the stimulatory G-protein, inhibitory G-proteins (Gi) exist. By activation of a different receptor and this Gi, adenylate cyclase is inhibited. In each category, different G-proteins have been identified (Gs1, Gs2, Gs3 and Gi1, Gi2, Gi3).

Phosphoinositide system - generates 2 second messengers, IP-3 and DAG. Upon receptor stimulation and G-protein activation, phospholipase C is stimulated, which hydrolases membrane phosphatidyl inositol 4,5-biphosphate into IP-3 and DAG. IP-3 releases calcium from intracellular stores, and DAG activates protein kinase C. Effects on ion channels or phosphorylation of specific proteins causes the physiologic effects.

Pharmacology of Neurotransmission

Drugs can increase or decrease the synthesis, storage, release, or degradation of a specific NT, usually in all neurons in the body and brain. Such drugs are not selective in their action on a specific organ; in contrast, certain drugs can rather selectively stimulate (agonists) or block (antagonists) specific receptors and can be more selective (concentration dependent) in their effects.

PHARMACOGENETICS

A pharmacogenetic reaction is a variation in drug response caused by hereditary factors. Many of these effects are unexpected and can be adverse. Pharmacogenetic responses may be classified as either direct or indirect.

Direct Pharmacogenetic Response

Reduced warfarin activity: Certain individuals exhibit lower anticoagulant activity following usual therapeutic doses of warfarin; a dose up to 20 times greater than normal may be necessary to produce the desired pharmacologic effect.

Malignant hyperthermia: After receiving a combination of muscle relaxant (usually succinylcholine) and inhalation general anesthetic (most frequently halothane), certain patients (about 1:20,000) undergo a life-threatening elevation in body temperature. Muscular rigidity is often the first sign; others (in addition to hyperthermia) include tachycardia and other arrhythmias, acidosis, and shock. The mechanism appears related to halothane-induced potentiation of Ca activity in skeletal muscle (sarcoplasmic reticulum); in susceptible patients, this tissue is hyperreactive to Ca.

Indirect Pharmacogenetic Response

Reduced biotransformation: Adverse reactions to certain drugs develop more frequently and at lower therapeutic doses in patients with this type of enzymatic alteration.

Acetylation: In about 50% of the US population, hepatic N-acetyltransferase is hypoactive (slow acetylators). As a result, isoniazid and certain other drugs (eg, hydralazine, phenelzine, procainamide, sulfamethazine) are slowly metabolized by this enzyme. Slow acetylators tend to be more susceptible to adverse reactions associated with these medications, eg, peripheral neuritis (isoniazid), lupus erythematosus (hydralazine, procainamide), sedation and nausea (phenelzine).

Hydrolysis: Persons with plasma pseudocholinesterase deficiency (about 1:2500) have a decreased ability to inactivate succinylcholine, resulting in prolonged paralysis of the respiratory muscles when conventional doses are administered. Prolonged apnea may require mechanical ventilation.

Oxidation: Reduced hydroxylation of various drugs has been correlated with an unusually large therapeutic response (eg, abnormally high degree of beta receptor blockade with metoprolol or timolol) or greater toxicity than expected (eg, excessive CNS depression with nortriptyline or phenytoin). Other drugs that appear to be affected by this metabolic difference include tricyclic antidepressants (eg, amitriptyline, desipramine) and an antitussive agent (dextromethorphan).

Accelerated biotransformation: Patients with an increased biotransformational capacity will require therapeutic doses larger than usual; they may also be more susceptible to certain toxic effects.

Acetylation: While almost half of the US population exhibits reduced N-acetyltransferase activity, the remainder are rapid acetylators. These patients require larger or more doses of isoniazid.

Oxidation: Alcohol dehydrogenase (ADH) is an important enzyme in the biotransformation of alcohol, oxidizing it to acetaldehyde. Apparently, about 85% of the Japanese population has an atypical ADH which operates about 5 times faster than normal; other Asian groups may exhibit this same phenomenon. Consumption of alcohol by such persons leads to accumulation of acetaldehyde, resulting in extensive vasodilation, facial flushing, and compensatory tachycardia.

Enzyme deficiency:

Glucose-6-phosphate dehydrogenase (G6PD): enzyme is essential for those RBC reactions which maintain cellular integrity. Patients with G6PD deficiency (including about 10% of blacks) are at increased risk of developing hemolytic anemia when given oxidant drugs such as antimalarials (eg, chloroquine, pamaquine, primaquine), aspirin, probenecid, or vitamin K.

Glutathione synthetase: is found in erythrocytes and hepatocytes. Similar to G6PD deficiency lack of RBC glutathione synthetase results in development of hemolytic anemia after administration of oxidant drugs.

KINETIC PRINCIPLES OF DRUG ADMINISTRATION

The study of the time course of a drug and its metabolites in the body following drug administration by any route.

Drugs are administered to achieve a therapeutic objective. This usually requires the attainment and maintenance of a pharmacologic response, which in turn requires an appropriate concentration of drug at the site of action. The appropriate concentration and the dosage needed to achieve it depend upon the patient's clinical state, the severity of the condition being treated, the presence of other drugs and concurrent disease, and other factors.

BASIC PHARMACOKINETIC PARAMETERS

Bioavailability and Absorption Rate Constant

The extent of drug absorption into the general circulation is expressed by the bioavailability, the fraction of a dose reaching the plasma site of measurement. The speed of absorption is often expressed by the absorption rate constant. Changes in these 2 parameters influence the maximum (or peak) concentration, the time at which the maximum concentration occurs, and the area under the concentration-time curve after a single oral dose. In chronic drug therapy, bioavailability is the more important measurement because it relates to the average level obtained, whereas the degree of fluctuation is related to the absorption rate constant.

29

Volume of Distribution and Unbound Fraction

The apparent volume of distribution and the fraction unbound in plasma are the 2 most widely used parameters for drug distribution. The volume of distribution is useful because it allows estimation of the dose required to achieve a given concentration and, conversely, the concentration achieved on administering a given dose. The unbound fraction is useful because it relates the measured total concentration to the unbound concentration, which is presumably more closely associated with drug effects. It is a particularly useful parameter when plasma protein binding is altered, eg, in hypoalbuminemia (a low albumin level), renal disease, hepatic disease, and displacement interactions.

Clearance, Renal Clearance, and Fraction Excreted Unchanged

The rate at which a drug is eliminated from the body is proportional to the plasma concentration. The parameters relating rates of renal excretion and metabolism to the plasma concentration are renal clearance, and metabolic clearance. Because the rate of elimination is the sum of the rates of renal excretion and extrarenal elimination, it follows that

Total clearance = Renal clearance + Extrarenal (metabolic) clearance

The ratio of the rate of renal excretion to the rate of total elimination, also the ratio of renal clearance to (total) clearance, is the fraction excreted unchanged. This parameter is useful in assessing the potential effect of renal and hepatic diseases on drug elimination.

Elimination Rate Constant and Half-life

The elimination rate constant relates the rate of elimination to the amount of drug in the body. As the rate of elimination equals clearance times plasma drug concentration and the amount of drug in the body equals volume of distribution times plasma drug concentration, it follows that

$$\text{Elimination rate constant} = \frac{\text{Clearance}}{\text{Volume of distribution}}$$

Expressed in these terms, the elimination rate constant is a function of how a drug is cleared from the blood by the eliminating organs and how the drug distributes throughout the body.

Half-life (elimination) is the time required for the plasma drug concentration or the amount in the body to decrease by 50%. For most drugs, the half-life remains constant regardless of how much drug is in the body. It is related to the elimination rate constant (0.693) by

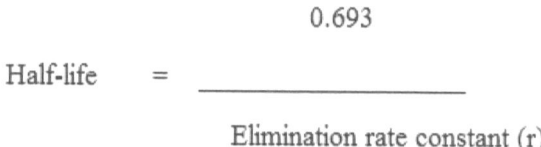

$$\text{Half-life} = \frac{0.693}{\text{Elimination rate constant (r)}}$$

VARIABILITY IN PARAMETER VALUES

Many of the variables affecting pharmacokinetic parameters have been recognized and can be taken into account

Age and Weight

For some drugs, changes in pharmacokinetics with age and weight are well established. In children and young people, renal function appears to correlate well with body surface area. Thus, for drugs primarily eliminated unchanged by renal excretion, clearance varies with age according to the change in surface area. In persons over age 20, renal function decreases about 1%/yr. For neonates and young infants, both renal and hepatic functions are not fully developed and no generalization, except for the occurrence of rapid change, can be made.

Disease

Renal function impairment: The renal clearance of most drugs appears to vary directly with creatinine clearance, regardless of the renal disease present. The change in (total) clearance depends upon the contribution of the kidneys to total elimination. Thus, (total) clearance is expected to be proportional to renal function (creatinine clearance) for drugs solely excreted unchanged and not to be affected for drugs eliminated by metabolism.

Hepatic disease produces changes in metabolic clearance, but good correlates or predictors of the changes are unavailable. Dramatically reduced drug metabolism has been associated with hepatic cirrhosis. Reduced plasma protein binding is often observed in this disease because of lowered plasma albumin. Acute hepatitis, with elevated serum enzymes, is usually not associated with altered drug metabolism. Heart failure, pneumonia, hyperthyroidism, and many other diseases also alter the pharmacokinetics of drugs.

Drug Interactions

Drug interactions can cause changes in pharmacokinetic parameter values and, therefore, in drug response. Most of these interactions are graded, and the extent of the interaction depends upon the concentrations of both of the interacting drugs.

Dose and Time Dependence

In some instances, the values of the pharmacokinetic parameters change with dose administered, concentration in plasma, or time; eg, a decreased bioavailability of griseofulvin as the dose is increased, a disproportionate increase in the steady-state plasma phenytoin concentration on increasing its dosing rate, and a decrease in plasma carbamazepine concentration during its chronic administration. The decreased bioavailability of griseofulvin is due to the drug's low solubility in the GI tract. Phenytoin shows a concentration (dose) dependency because the metabolizing enzymes have a limited capacity to eliminate the drug, and the usual rate of administration approaches the maximum rate of metabolism. Carbamazepine shows time dependence because it induces its own metabolism.

MONITORING DRUG TREATMENT

Once a therapeutic objective is defined and a drug and dosage regimen are chosen for a patient, drug therapy is conventionally managed by monitoring the incidence and intensity of both therapeutic and undesirable effects.

MONITORING DRUG IN PLASMA

Plasma drug concentration monitoring is a procedure that can provide a facile and rapid estimation of dosage requirements. For some drugs it is routinely useful; for others it can be helpful in certain situations.

INDICATIONS FOR MONITORING

Criteria Related to the Drug

1. The intensity and probability of therapeutic or toxic effects must correlate quantitatively with the plasma level.

2. The objective of the regimen must be to attain and maintain a therapeutic effect.

3. When assessed therapeutic end points are lacking, plasma concentration monitoring becomes particularly attractive; e.g., in antiepileptic therapy, for which the therapeutic end point is the absence of seizures.

4. The probability of a therapeutic problem is greater for a drug with a low margin of safety or a low therapeutic index.

5. Prior knowledge of the therapeutic concentrations and the pharmacokinetic parameters of a drug is essential for plasma concentration monitoring to be effective.

6. Inter-individual differences and, in certain conditions, intra-individual differences in the pharmacokinetics of drugs are principal reasons for monitoring plasma concentrations.

Criteria Related to the Situation

For a patient with GI disease or with a gastric resection, an orally administered drug known to have poor bioavailability may be a candidate for monitoring. Similarly, the presence of renal, hepatic, thyroid, or cardiovascular disease may also suggest monitoring. For drugs that are primarily excreted unchanged, the presence of renal disease requires special attention.

THE THERAPEUTIC WINDOW

The range of plasma concentrations with a high probability of therapeutic success.

For drugs that are bound to plasma proteins and in situations in which an alteration in binding is anticipated, the total concentration (bound + unbound) must be adjusted to give the desired unbound concentration. Conditions that reduce binding to albumin include end-stage renal disease, cirrhosis, hypoalbuminemia (low levels of albumin), severe burns, and pregnancy. Binding to alpha1-acid glycoprotein and lipoproteins has been observed to be increased during stress and decreased in chronic hepatic disease. In these cases, adjustment of the therapeutic window is accomplished by estimating the fraction unbound in plasma in the patient and comparing it to the usual fraction unbound. Thus,

$$\text{Adjusted concentration} = \frac{\text{Usual fraction unbound}}{\text{Anticipated fraction unbound}} \times \text{Usual concentration}$$

EVALUATION OF A MEASURED CONCENTRATION

Steady state: A value that represents an estimate of the average steady-state concentration on a fixed-dose, fixed-dosing-interval regimen is handled most readily. This requires a plasma sample obtained after dosing for at least 3 half-lives.

The predicted average concentration (Cav) is a function of the expected values of bioavailability (F), clearance (CL), and the rate of administration (D/tau, dose/dosing interval), that is,

$$\text{Cav (expected)} = \text{(expected)} \frac{F}{CL} \times \frac{D}{tau}$$

If the ratio of concentrations, observed to predicted, is >1, either the input is greater or the elimination is slower than expected, or both. The converse is true for a ratio <1.

When absorption and distribution are rapid, measurement of plasma concentration soon after the dose and close to the peak has been found to be useful. Under these conditions, the peak concentration can be estimated from

$$\text{Cpeak} = \frac{F \times D}{V\left(1 - e^{-\left[\frac{CL}{V} \times tau\right]}\right)}$$

or from the relationship

$$\mathrm{Cpeak} \;=\; C \,+\, F \times D/V$$

where $F \times D/V$ is the increment of change in the concentration on adding $F \times D$ to the body.

Estimation of Parameter Values

For adjusting dosage in an individual patient, the most useful procedure is to estimate the value of clearance and sometimes the values of the volume of distribution and half-life from the monitored concentrations. Clearance is the most valuable parameter, because it is needed to predict the dosage required to achieve a given concentration and the converse.

From a steady-state value:

$$CL \quad = \quad F \times D/tau$$

$$\text{---------------}$$

$$Cav$$

Clearance may be calculated from the relationship:

$$CL \quad = \quad \frac{V}{tau} \quad \times \quad \ln \left[\frac{F \times D}{V \times C} + 1 \right]$$

The dosage required to achieve a given average concentration can then be computed from the clearance and bioavailability estimates as follows:

$$D/tau \quad = \quad \frac{CL}{F} \times Cav$$

To calculate the loading dose for achievement of required concentration -

$$L d. = V \times C / S \times F$$

where S is an active part of the drug, C – required concentration, V – distribution volume, F – bioavailability.

To calculate the maintenance dose for achievement of required concentration -

$$M d. = CL \times C / F$$

To calculate the distribution volume -

$$V = CL / r \text{ (elimination constant}$$

$= 0.7)$

List of commonly monitored drugs

Aminoglycosides
Phenitoin
Phenobarbital
Digoxin
Coumadine
Heparin
Theophylline
Quinidine
Procainamide
Tobramycine
Vancomycine

DRUG INTERACTIONS AND METABOLISM

Alteration of the effects of one drug by the prior or concurrent administration of another (drug-drug interactions); alteration of the effects of a drug by food (drug-food interactions). The effects of one of the drugs are usually increased or decreased. Desired interactions are achieved with combination in which 2 or more drugs are used to increase therapeutic effects or reduce toxicity. Unwanted interactions can cause adverse drug reactions or therapeutic failure.

Since it is often difficult to predict the clinical significance of known or suggested drug interactions, the possibility of problems developing must be viewed in perspective. If an interaction appears likely, therapeutic alternatives should be considered; but a patient should not be denied needed therapy solely because of the possibility.

The mechanisms of drug interactions are usually pharmacodynamic or pharmacokinetic.

PHARMACODYNAMIC INTERACTIONS

Pharmacodynamic interactions include the concurrent administration of drugs having the same (or opposing) pharmacologic actions and alteration of the sensitivity or the responsiveness of the tissues to one drug by another. Many of these interactions can be predicted from knowledge of the pharmacology of each drug

Drugs with Opposing Pharmacologic Effects

Interactions resulting from the use of 2 drugs with opposing pharmacologic effects should be among the easiest to detect, but various factors may preclude early identification of such antagonism. For example, thiazides and certain other diuretics may elevate blood glucose levels. When a diuretic is prescribed for a diabetic who takes insulin or an oral hypoglycemic agent, the hypoglycemic action of the antidiabetic drug may be partially counteracted, necessitating a dosage adjustment.

Drugs with Similar Pharmacologic Effects

An example of this type of interaction is the increased CNS-depressant effect that often occurs when persons taking antianxiety agents, antipsychotic agents, antihistamines, or other drugs having depressant effects drink alcoholic beverages. Many people risk these combinations without serious difficulty, but they can be lethal. Combining drugs with CNS-depressant activity increases the risks of excessive sedation and dizziness.

Excessive anticholinergic effects are common with the concurrent use of drugs such as an antipsychotic agent (e.g., chlorpromazine), an antiparkinsonian drug (e.g., trihexyphenidyl), and a tricyclic antidepressant (e.g., amitriptyline). In some individuals, particularly elderly patients, this additive effect may result in an atropine-like delirium that could be mistakenly interpreted as a worsening of psychiatric symptoms or the presence of dementia. Concurrent use of drugs with anticholinergic activity may also result in dry mouth and associated dental complications, blurring of vision, and hyperpyrexia in patients exposed to high temperature and humidity.

Commonly, a patient may unknowingly take several different products that contain the same nonsteroidal anti-inflammatory drug.

An arthritic patient, using prescribed ibuprofen (often at dosage levels at or near the recommended maximum), may purchase an OTC ibuprofen product for pain or discomfort not associated with the arthritis, not knowing that the 2 products contain the same drug and increase the risk of adverse effects.

Interactions at Receptor Sites

The enzyme monoamine oxidase (MAO) metabolizes catecholamines such as norepinephrine. Norepinephrine accumulates within the adrenergic neurons when MAO is inhibited. Drugs that cause a release of these greater-than-usual amounts of norepinephrine can bring about exaggerated responses, including severe headache, hypertension (possibly a hypertensive crisis), and cardiac arrhythmias. Such an interaction may occur between MAO inhibitors (e.g., isocarboxazid, phenelzine, tranylcypromine, pargyline) and indirectly acting sympathomimetic amines. While most sympathomimetic amines (e.g., amphetamine) are available only by prescription, others (e.g., ephedrine, phenylephrine, phenylpropanolamine), known to interact with MAO inhibitors, are present in many popular OTC cold, allergy, and diet remedies. Patients taking MAO inhibitors should avoid using such products.

Serious reactions (hypertensive crises) have occurred in patients being treated with MAO inhibitors following the ingestion of foods and beverages having a high tyramine content, including certain cheeses, alcoholic beverages, concentrated yeast extracts, broad-bean pods, and pickled herring. Tyramine is metabolized by MAO, which is present in the intestinal wall and the liver; this enzyme protects against the pressor actions of amines in foods. When the enzyme is inhibited, unmetabolized tyramine can accumulate, releasing norepinephrine from adrenergic neurons.

The antineoplastic drug procarbazine and the anti-infective drug furazolidone (or probably its metabolite) can also inhibit MAO, and the same warnings apply to these drugs as to other MAO inhibitors.

PHARMACOKINETIC INTERACTIONS

Pharmacokinetic interactions are more complicated and difficult to predict because the interacting drugs often have unrelated actions; the interactions are mainly due to alteration of absorption, distribution, metabolism, or excretion, which changes the amount and duration of a drug's availability at receptor sites.

Alteration of Gastrointestinal Absorption

Interactions that involve a change in drug absorption from the GI tract are of variable importance.

Alteration of pH: Many drugs are weak acids or weak bases, and the pH of the GI contents can influence absorption. Since the nonionized (more lipid-soluble) form of a drug is more readily absorbed than the ionized form, acidic drugs are usually more readily absorbed from the upper regions of the GI tract, where they are primarily in a nonionized form.

Complexation and adsorption: Tetracyclines can combine with metal ions (e.g., Ca, Mg, Al, and Fe) in the GI tract to form poorly absorbed complexes. Thus, certain foods (e.g., milk) or drugs (e.g., antacids, products containing Mg, Al, and Ca salts, or Fe preparations) can significantly decrease tetracycline absorption. The increase in pH of the GI contents probably also contributes to the reduction of the tetracycline absorption.

Antacids markedly reduce the absorption of fluoroquinolone derivatives (e.g., ciprofloxacin), probably as a result of the metal ions complexing with the drug. Antacids should not be used simultaneously or <2 h (or preferably, an even longer period) after ciprofloxacin.

Complexation can be expected with cholestyramine and colestipol. In addition to binding with and preventing reabsorption of bile acids, these agents can bind with drugs in the GI tract, having the greatest affinity for acidic drugs, e.g., thyroid hormone or warfarin. To minimize the possibility of such an interaction, the interval between taking cholestyramine or colestipol and another drug should be as long as possible

Some antidiarrheals (e.g., those containing kaolin and pectin), may adsorb other drugs, resulting in decreased absorption.

Alteration of motility: By increasing GI motility, metoclopramide may hasten the passage of drugs through the GI tract, resulting in decreased absorption, particularly of drugs that require prolonged contact with the absorbing surface and those that are absorbed only at a particular site along the GI tract. Similar problems can occur with enteric-coated and sustained-release formulations.

By decreasing GI motility, anticholinergics may either reduce absorption by retarding dissolution and slowing gastric emptying, or increase absorption by keeping a drug for a longer period of time in the area of optimal absorption.

Effect of food: Food may delay or reduce the absorption of many drugs. Food often slows gastric emptying, but it may also affect absorption by binding with drugs, by decreasing their access to absorption sites, by altering their dissolution rates, or by altering the pH of the GI contents.

Food in the GI tract will reduce the absorption of many antibiotics. Although there are exceptions (e.g., penicillin V potassium, amoxicillin, doxycycline, minocycline), it is generally recommended that penicillin and tetracycline derivatives, erythromycin stearate and formulations of erythromycin base that are not enteric coated, as well as several other antibiotics, be given at least 1 h before meals or 2 h after meals to achieve optimal absorption. Food has also been reported to decrease the absorption of many other therapeutic agents including astemizole, captopril, and penicillamine.

Alteration of Distribution

Displacement of drugs from protein-binding sites may occur when 2 drugs capable of protein binding are given concurrently, especially when they are capable of binding to the same sites on the protein molecule (competitive displacement). Since the number of plasma or tissue protein-binding sites is limited, drugs can displace one another.

Both phenylbutazone and warfarin are extensively bound to plasma proteins, especially albumin, but phenylbutazone has a greater affinity for the binding sites. When the 2 drugs are taken concurrently, fewer binding sites are available for warfarin, thus increasing the amount of free anticoagulant and the risk of hemorrhage. Phenylbutazone also inhibits the metabolism of warfarin, resulting in continued enhancement of its anticoagulant effect.

Alteration of Metabolism

Stimulation of metabolism: Many drug interactions result from the ability of one drug to stimulate the metabolism of another by increasing the activity of hepatic enzymes involved in their metabolism (enzyme induction). In this manner, phenobarbital increases the rate of metabolism of coumarin anticoagulants such as warfarin, resulting in a decreased anticoagulant response. Phenobarbital also accelerates the metabolism of other drugs such as steroid hormones. Enzyme induction also is caused by other barbiturates and by various therapeutic agents (e.g., carbamazepine, phenytoin, and rifampin).

Disturbed calcium metabolism and osteomalacia are associated with the use of anticonvulsants such as phenobarbital and phenytoin. Reduced serum calcium levels are caused by vitamin D deficiency, resulting from enzyme induction by the anticonvulsants. Pyridoxine can antagonize the activity of the antiparkinsonian drug levodopa by accelerating the conversion of the levodopa to its active metabolite, dopamine, in the peripheral tissues. In contrast to levodopa, dopamine cannot cross the blood-brain barrier, where it is required for the antiparkinsonian effect. In patients receiving both levodopa and carbidopa (a decarboxylase inhibitor), the addition of pyridoxine does not reduce the action of levodopa.

Efficacy of certain drugs (e.g., chlorpromazine, diazepam, propoxyphene, theophylline) may be decreased in individuals who smoke heavily, because of increased hepatic enzyme activity from the action of polycyclic hydrocarbons found in cigarette smoke.

Drugs causing induction of hepatic mitochondrial enzymes(P-450)

Barbiturates, rifampin, digoxin, phenytoin (decreasing levels of – steroids, theophylline, warfarin, quinine).

Inhibition of metabolism: One drug may inhibit the metabolism of another, causing its prolonged and intensified activity. For example, disulfiram, used in the treatment of alcoholism, inhibits the activity of aldehyde dehydrogenase, thus inhibiting the oxidation of acetaldehyde, an oxidation product of alcohol. This results in the accumulation of excessive acetaldehyde and causes the characteristic disulfiram effect following the consumption of alcohol. Disulfiram also enhances the activity of warfarin and phenytoin by inhibiting their metabolism.

Allopurinol reduces the production of uric acid by inhibiting the enzyme xanthine oxidase. However, xanthine oxidase is involved in the metabolism of such potentially toxic drugs as mercaptopurine and azathioprine; when the enzyme is inhibited, the effect of these 2 agents can be markedly increased. Therefore, when allopurinol is given concurrently, a reduction to about 1/3 to 1/4 the usual dose of mercaptopurine or azathioprine is advised.

Cimetidine inhibits oxidative metabolic pathways and is likely to increase the action of other drugs that are metabolized via this mechanism (e.g., carbamazepine, phenytoin, theophylline, warfarin, and certain benzodiazepines). Most benzodiazepines (e.g., diazepam) are metabolized via oxidative mechanisms; however, lorazepam, oxazepam, and temazepam undergo glucuronide conjugation and their action is not affected by cimetidine. Although ranitidine also binds to the hepatic oxidative enzymes, it appears to have less affinity for the enzymes than does cimetidine,famotidine and nizatidine are not inhibit oxidative metabolic pathways.

Erythromycin inhibit the hepatic metabolism of agents such as carbamazepine and theophylline, thereby increasing their effects. The fluoroquinolones (e.g., ciprofloxacin) also increase the activity of theophylline, presumably by the same mechanism.

Drugs causing inhibition of hepatic mitochondrial enzymes (P-450)

Isoniazide, cimetidine, allopurinol, disulfiram, TCA, oral contraceptive, erythromycin, methotrexate, chloramphenicol (increase levels of tolbutamide, phenytoin, theophylline, benzodiazepines, barbiturates).

Alteration of Urinary Excretion

Alteration of urinary pH: Urinary pH influences the ionization of weak acids and bases and thus affects their reabsorption and excretion. A nonionized drug more readily diffuses from the glomerular filtrate into the blood. More of an acidic drug is nonionized in an acid urine than in an alkaline urine, where it primarily exists as an ionized salt. Thus, more of an acidic drug (e.g., a salicylate) diffuses back into the blood from an acid urine, resulting in prolonged and perhaps intensified activity. The risk of a significant interaction is greatest in patients who are taking large doses of salicylates (e.g., for arthritis). Opposite effects are seen for a basic drug like dextroamphetamine.

Alteration of active transport: Probenecid increases the serum levels and prolongs the activity of penicillin derivatives, primarily by blocking their tubular secretion. Such combinations have been used to therapeutic advantage.

Significantly greater serum digoxin levels are found when quinidine is administered concurrently than when digoxin is given alone. Quinidine appears to reduce the renal clearance of digoxin, although other nonrenal mechanisms are probably also involved in this interaction.

A number of nonsteroidal anti-inflammatory drugs (NSAIDs) increase the activity and toxicity of methotrexate. There have been reports of fatal methotrexate toxicity in patients also receiving ketoprofen, and it has been suggested that ketoprofen inhibited the active renal tubular secretion of methotrexate.

DRUGS ACTING ON THE SYMPATHETIC NERVOUS SYSTEM

Background

Catecholamines are released by the adrenal gland and by the sympathetic nervous system. Their main function is adapting the body to deal with stressful situations. The endogenous catecholamines are epinephrine (from the adrenal), norepinephrine (in the sympathetic nervous system) and dopamine (in the CNS).

A.   ADRENERGIC AGONISTS – CATECHOLAMINES AND SYMPATHOMIMETIC DRUGS

Background

1.   There are two main classes of receptors – a and b, which are further divided into a1, a2, b1, b2 and b3. Different drugs have different effects on the various receptors.

2.   b1 receptors are found in heart muscle.

3.   b2 receptors are found in bronchial smooth muscle and blood vessels of skeletal muscle.

4.   a receptors are in blood vessel walls.

Epinephrine

1.   acts on both a and b receptors

2.   Effects

a.   Cardiovascular

(1) Potent elevator of blood pressure. Intravenous injection causes dose-dependent increases in blood pressure; systolic pressure is more affected than diastolic and so pulse pressure is increased. Effects are due to increased heart rate, increased contractions of the myocardium and vasoconstriction. Subcutaneous administration or intravenous infusion causes a lower increase in systolic blood pressure with decreased peripheral vascular resistance and diastolic blood pressure.

(2) Vasoconstriction of subcutaneous small vessels, increased blood flow to skeletal muscle

(3) Stimulation of myocardium – increased pulse, arrhythmias, increased work and oxygen consumption, increased cardiac output

b.   Smooth muscle – increased blood flow, relaxation of GIT and bladder muscle, b2 selective agonists relax uterine smooth muscle

c.   Respiratory system – relaxes bronchi

d.   Metabolic effects – increased oxygen consumption, hyperglycemia, lactic acidosis, increased free fatty acids (b receptors); effects on insulin depend on receptor – a2 inhibit secretion, b2 stimulate it, the result is inhibition. A transient hyperkalemia is followed by a more persistent hypokalemia.

e.   Other effects – eisonopenia (a decrease in the number of eosinophils in the blood), increased coagulability of blood, decreased intraocular pressure, mydriasis (Dilatation of the pupil), tears

3.   When given orally, too quickly broken down to be of clinical use. Subcutaneous is slow, due to vasoconstriction. Intramuscular is the most effective route of administration. When used in inhalers, the concentrations are good in the respiratory tract, with minimal (but existing) systemic side effects.

4.   Uses

a.   Bronchodilatation in asthma and to relieve bronchospasm

b.   To treat hypersensitivity reactions and anaphylactic shock

c.   Used together with local anesthetics to prolong duration by causing vasoconstriction

d.   Used in cardiac arrest

5.   Toxic effects

a.   May be transient – fear, anxiety, dizziness, pallor, tremors, headache and palpitations

b.   More serious effects are arrhythmias and cerebral hemorrhage – due to rapid elevation of blood pressure

c.   Worse in patients with psychiatric backgrounds, hypertension or hyperthyroidism

6.   Contraindications – patients receiving nonselective beta blockers (a effects unopposed), patients with both emphysema and heart disease

Norepinephrine

1.   More potent than epinephrine on a receptors, much less potent on b2 receptors and the same on b1 receptors

2.   Effects

a.   Cardiovascular

(1)   Increases in systolic, diastolic and pulse pressures; increased peripheral vascular resistance, but no change in cardiac output

(2) Sinus bradycardia, arrhythmias

b.   Other effects, as seen with epinephrine, are seen only at high doses of norepinephrine

3.   Routes of administration as for epinephrine

4.   Uses – used in shock

5.   Toxic effects – as with epinephrine, but milder

a.   Severe hypertension, headache, photophobia, pallor, sweating and vomiting

b.   Increased risk of arrhythmias

6.   Contraindication – pregnancy (will cause uterine contractions)

Dopamine

1.   Precursor of both epinephrine and norepinephrine

2.   Effects

a.   CNS effects are minimal when given intravenously; does not cross blood-brain barrier

b.   Cardiovascular effects are dose dependent.

(1)   The first effects are vasodilatation, increased renal blood flow and glomerular filtration rate (GFR) – mediated through dopamine receptors.

(2)   Increased dose has positive isotropic (Affecting the force or energy of muscular contractions) effects with increased systolic and pulse pressure (slight effect on diastolic pressure) and no change in total peripheral resistance. These are mediated through b1 receptors.

(3)   With higher dose, vasoconstriction results, mediated through a1 receptors.

3.   Not effective orally as it is rapidly broken down; used intravenously only

4.   Uses – treatment of shock. Lowest dose is used to treat oliguria in hydrated patient.

5.   Side effects – nausea, vomiting, tachycardia, chest pain, headache, hypertension, vasoconstriction and arrhythmias

6.   Contraindications – (relative) patients receiving monoamine oxidase inhibitors

Amphetamines

1.   Other similar drugs are methylphenidate, ephedrine, pemoline and methamphetamine

2.   Effects

a.   Increased systolic and diastolic blood pressure

b.   Contraction of bladder

c.   CNS stimulation – alertness, lack of fatigue, euphoria, self-confidence, increased concentration, enhanced physical performance

d.   Compensates for lack of sleep

e.   Depresses appetite (but is tolerance)

3.   Uses – obesity, narcolepsy and attention deficit hyperactivity disorder (methylphenidate)

4.   Can be given orally

5.   Side effects – increased errors in tasks performed, headache, palpitations, arrhythmias, depression, fatigue, dry mouth, increased sweating, nausea and vomiting, abdominal pain, confusion and psychomotor agitation.

6.   Chronic use may cause psychotic reactions.

7.   Toxic doses can cause convulsions, coma and death.

B.   SELECTIVE ADRENERGIC AGONISTS

Methoxamine

1.   a1 selective agonist

2.   Effects – increased peripheral vascular resistance, elevated blood pressure, sinus bradycardia

3.   Uses – in hypotension, shock and paroxysmal atrial tachycardia

4.   Given intravenously

Phenylephrine

1.   a1 selective agonist

2.   Effects – increased peripheral vascular resistance, elevated blood pressure, sinus bradycardia (Slow heart rate), vasoconstriction

3.   Uses – as nasal decongestant and to dilate pupils

4.   Given intravenously and topically to the nose and eyes

5.   Side effects – strong vasoconstriction when given intravenously

Clonidine

1.   a2 selective agonist

2.   Effects – vasoconstriction, hypotension (intravenous administration causes transient hypertension with prolonged hypotension; oral causes hypotension only), bradycardia (Slow heart rate) and sedation

3.   Uses – main use is as antihypertensive; other uses are in treatment of substance addiction, in the relief of vasomotor symptoms of the menopause and in anesthesia

4.   Can be given orally or as transdermal patch

5.   Side effects – dry mouth, sedation are very common; less so are sexual dysfunction and serious bradycardia.   Patches can cause contact dermatitis.

Methyldopa

1.   a2 selective agonist

2.   Centrally acting pro-drug

3.   Effects – reduces peripheral resistance with normal renal blood flow

4.   Given orally or intravenously

5.   Uses – antihypertensive, can be used in pregnant women, especially useful in left ventricular hypertrophy

6.   Side effects - mild and transient sedation, dry mouth, reduced libido, Parkinsonism, hyperprolactinemia

7.   Toxic effects

a.   hepatitis – usually transient, can be fatal

b.   hemolytic anemia

c.   rarer – leukopenia, thrombocytopenia, SLE, myocarditis, retroperitoneal fibrosis, pancreatitis

Isoproterenol

1.   b selective agonist

2.   Effects

a.   Decreased peripheral vascular resistance and diastolic pressure (systolic pressure usually unchanged), increased cardiac output due to positive inotropic and positive chronotropic effects

b.   Relaxes smooth muscle of respiratory and GI tracts

c.   Mild hyperglycemia

3.   Can be inhaled or given orally

4.   Uses – to increase heart rate in bradycardia or heart blocks

5.   Side effects – palpitations, sinus tachycardia , ischemia, arrhythmias, flushing, headache

Dobutamine

1.   b selective agonist

2.   Effects – more inotropic than chronotropic, increases stroke volume and cardiac output

3.   Given intravenously

4.   Used in congestive heart failure, acute myocardial infarction and before cardiac surgery

5.   Side effects – hypertension, tachycardia

Metaproterenol

1.   b2 selective agonist

2.   Effects - bronchodilatation

3.   Given orally and as inhalant (good absorption with minimal systemic side effects)

4.   Uses – acute bronchospasm, asthma, Chronic Obstructive Pulmonary Disease (COPD)

5.   Side effects – tremor (tolerance later develops), apprehension, anxiety,

Terbutaline

1.   b2 selective agonist

2.   Effects - bronchodilatation

3.   Given orally, subcutaneously and as inhalant

4.   Uses - acute bronchospasm, asthma, COPD and for status asthmaticus; also used to treat uterine contractions

Ritodrine

1.   b2 selective agonist

2.   Effects – relaxation of uterine muscle

3.   Given orally and intravenously

4.   Uses – to treat premature uterine contractions

5.   Toxic effects – pulmonary edema, worsening of cardiac disease

C.   ALPHA ADRENERGIC ANTAGONISTS

Phenoxybenzamine

1.   Irreversible blockade of both a receptor types

2.   Effects – decreased peripheral vascular resistance, increased cardiac output, tachycardia

3.　　　Used in the treatment of pheochromocytoma to cause a blockade and in benign prostatic hyperplasia in poor surgical risks.

4.　　　Side effects - orthostatic hypotension, reflex tachycardia , arrhythmias, inhibition of ejaculation (reversible)

Prazocin

1.　　　Selective a1 antagonist

2.　　　Effects – vasodilatation, decreased peripheral resistance and hypotension, without reflex tachycardia. Both preload and afterload are reduced.

3.　　　Can be given orally

4.　　　Used in the treatment of hypertension, congestive heart failure and benign prostatic hyperplasia

5.　　　Side effects – syncope (with the first dose) – start at low doses, best to give at bedtime

E. BETA ADRENERGIC ANTAGONISTS (BETA BLOCKERS)

General information about beta blockers

1.　　　Effects

a.   Cardiovascular – negative inotropic and chronotropic effects (especially in cases of sympathetic stimulation; less prominent in normal situations)

(1)   Short-term administration causes decreased cardiac output and increased peripheral vascular resistance.

(2)   Long term use causes return of peripheral vascular resistance to normal.

(3)   Also regulate heart rate and rhythm by slowing sinus rate and conduction

b.   No effect on blood pressure in normotensive individuals; decreases blood pressure in hypertensive patients.

c.   Bronchoconstriction - no clinical effect in healthy people; can severely worsen situation in asthmatics or patients with COPD

2.   Uses

a.   Hypertension

b.   ischemic heart disease – including secondary prevention after myocardial infarction

c.   Supraventricular and ventricular arrhythmias

d.   Hypertrophic obstructive cardiomyopathy (idiopathic hypertrophic subaortic stenosis)

e.   Acute dissection of aortic aneurysm

f.   To relieve symptoms of hyperthyroidism (especially propanolol)

g.   Prophylaxis of migraine (propanolol, timolol, metoprolol)

h.   Glaucoma

3.   Side effects

a.   Cardiovascular – bradycardia, intermittent claudication

b.   Respiratory – bronchospasm (is less in b1 selective antagonists, but even they should be avoided in asthmatics)

c.   Metabolism – increased triglycerides, may delay recovery from hypoglycemia in diabetics and mask signs of hypoglycemia in diabetics (b1 selective antagonists may be better for diabetics)

d.   Serious side effects include inducing or worsening congestive heart failure in patients with heart disease, causing bradyarrhythmias in patients with conduction defects

4.   Toxic effects – hypotension, bradycardia, slowed AV node conduction, wide QRS complexes, seizures, depression

5.   Contraindications (relative) - do not use in patients with COPD or asthma (bronchoconstriction) or in diabetics (masks signs of hypoglycemia)

6.   Interactions

a.   Verapamil or antiarrhythmic agents may decrease conduction and increase chance of bradyarrhythmias.

b.   Decreased absorption with aluminum salts, cholestyramine, colestipol

c.   Decreased concentration with phenytoin, rifampin, phenobarbital, tobacco use

d.   Increased bioavailability with cimetidine, hydralazine

e.   Beta-blockers decrease clearance of lidocaine.

f.   NSAIDs oppose antihypertensive effects of beta-blockers.

7.   Need to decrease dose gradually - abrupt cessation can worsen ischemic heart disease or cause sudden death

Propanolol

1.   Non-selective b antagonist with no agonist or a activity

2.   Can be given orally in regular or sustained release form, or intravenously

3.   Uses – hypertension, ischemic heart disease, prophylaxis of migraine headaches, hyperthyroidism (other than beta blockade, propanolol also inhibits peripheral conversion of T4 to T3), essential tremor

Timolol

1.   Non selective b antagonist

2.   Oral and topical to the eye

3.   Uses – glaucoma

Labetalol

1.   Blocks both a and b receptors, but more potent against b receptors

2.   Effects – hypotension, vasodilatation, decreased peripheral resistance

3.   Can be given orally and intravenously

4.   Uses - hypertension

Metoprolol

1.   b1 selective antagonist

2.   Given orally

3.   Uses – hypertension, stable angina pectoris, acute myocardial infarction

4.   Contraindications– do not use in acute myocardial infarction if pulse is less than 45 bpm, systolic blood pressure less than 100, heart failure is moderate or severe or P-R interval is 0.24 sec or more (2nd or 3rd degree block)

Esmolol

1.   b1 selective antagonist

2.   Very short acting

3.   Used in emergencies where short beta blockade is needed

Other common non-selective beta blockers are nadolol and pindolol. Other common b1 selective blockers are atenolol and acebutolol.

 # ANTIBIOTICS

Background

Divided by mode of action

1.   inhibit cell wall synthesis – penicillins, cephalosporins, carbapenems, vancomycin, bacitracin, miconazole, ketoconazole, clotrimazole

2.   increase permeability of cell membrane – polymyxin, nystatin, amphotericin B

3.   reversibly inhibit protein synthesis by affecting ribosomes 30 S and 50 S (bacteriostatic) – chloramphenicol, tetracyclines, macrolides, clindamycin

4.   alter protein synthesis via 30 S ribosomes (bactericidal) – aminoglycosides

5.   inhibit nucleic acids – rifampin, quinolones

6.   antimetabolites – trimethoprim, sulfonamides

7.   nucleic acid analogs – act on viruses – acyclovir, zidovudine, ganciclovir

PENICILLINS (Beta Lactams)

General side-effects

1.   Most important are hypersensitivity and anaphylaxis. Penicillins are the most common cause of drug allergy. All types, dosages and forms of administration can potentially cause allergic reactions. Cross-reactions among the penicillins and between them and the cephalosporins are not uncommon.

2.   Local reactions to intramuscular injection may be seen. Oral administration may cause nausea and diarrhea.

3.   Toxic effects are minimal.

Penicillin G and V

1.   effective against

a.   Gram-positive cocci (Streptococcus, Neisseria meningitidis, but not Enterococcus)

b.   Trepenoma pallidum (syphilis)

c.   Borrelia burgdorferi (Lyme disease)

d.   some anaerobes (Clostridium, Corynebacterium diphtheriae, Actinomyces, Bacillus anthracis, Listeria) – not Bacteroides fragilis

2.   Susceptible to penicillinase and so not effective against Staphylococcus aureus

3.   Given orally (penicillin V less vulnerable to gastric acid) or parenterally; adding procaine or benzathine prolongs the effect of the penicillin. Best absorption is on an empty stomach.

4.   Uses

a.   To treat infections by above bacteria

b.   As prophylaxis to prevent recurrent rheumatic fever, and in contacts of patients with syphilis, gonorrhea and streptococcal infections

c.   Used as prophylaxis for infective endocarditis for susceptible people before dental or surgical procedures

5.   Interactions – probenecid causes increased concentration

6.   Jarisch-Herxheimer reaction

a.   Seen in the majority of patients with secondary syphilis on receiving the first dose of penicillin

b.   Flu-like symptoms – chills, fever, muscle aches, headache, joint pain – with enhanced color of the lesions of syphilis

c.   Lasts for up to 48 hours and does not recur with later doses

d.   May be due to antigens released by breakdown of the spirochetes

e.   Not a reason to stop penicillin treatment; aspirin will help with symptoms

Penicillinase-resistant penicillins

1.   Examples are methicillin, nafcillin, oxacillin, cloxacillin, and dicloxacillin

2.   Effective against Gram-positive cocci (less so than penicillin V or G), including Staphylococcus aureus

3.   MRSA = methicillin resistant S. aureus are treated by vancomycin.

4.   Given orally (best absorption is on an empty stomach) and parenterally

5.   Methicillin may cause interstitial nephritis.

Ampicillin, etc.

1. Examples are ampicillin, amoxicillin, bacampicillin

2. Effective also against some Gram-negative organisms – i.e. Hemophilus, E. coli, Proteus, also Salmonella and Listeria

3. Susceptible to b-lactamase – and so not effective for Staph. infections

4. Given orally (best is on an empty stomach) and parenterally

5. Ampicillin causes a rash in patients with infectious mononucleosis. Allopurinol will also increase chances of a rash with ampicillin.

Ampicillin and clavulanic acid

1. Clavulanic acid inactivates b-lactamase and so prevents breakdown of the antibiotic

2. Thus, the compound is effective against Staph. organisms.

Anti-Pseudomonas penicillins

1.   Examples are carbenicillin and ticarcillin

2.   Effective also against Pseudomonas, Enterobacter and Proteus

3.   Susceptible to b-lactamase

4.   Carbenicillin increases risk of heart failure, due to its ability to cause hypernatremia and hypokalemia.   It also interferes with aggregation of platelets. Ticarcillin has fewer side effects and so is preferred; it is given orally and parenterally.

5.   A preparation of ticarcillin and clavulanic acid (for parenteral use) is more effective against Bacteroides, S. aureus and Gram-negative bacilli.

Broad-spectrum penicillins

1.   Examples are mezlocillin and piperacillin

2.   Effective also against Klebsiella and some other Gram-negative organisms

3.   Susceptible to b-lactamase

4.   Given only parenterally

CEPHALOSPORINS

General information

1.   Most are given orally, intramuscularly or intravenously. Cephalothin is painful when injected intramuscularly and so should be used intravenously only.

2.   Need to reduce dose in cases of kidney dysfunction

3.   Cefuroxime, moxalactam, cefotaxime, ceftriaxone, cefepime and ceftizoxime are found in therapeutic concentrations in the Cerebrospinal Fluid (CSF).

4.   Cephalosporins cross the placenta.

5.   Uses

a.   As treatment for bacteria as given below

b.   Prophylaxis for surgery

6.   Side effects

a.   Hypersensitivity reactions, as with penicillins; is cross-reactivity with penicillins

b.   Diarrhea

7.   Toxic effects – renal damage including acute tubular necrosis

8.   Interactions -   risk of nephrotoxicity increased with aminoglycosides

First generation

1.   Examples are cephalothin, cefazolin, cephalexin and cefadroxil. The first two are given parenterally, the second two orally.

2.   Effective against Gram-positive organisms (not including MRSA, S. epidermidis or Enterococcus) with slight activity against Gram-negative organisms (E. coli, Moraxella, Klebsiella, Proteus)

Second generation

1.   Examples are cefoxitin, cefotetan, cefmanadole, cefuroxime, cefonicid and ceforanide that are given parenterally and cefaclor and loracarbef that can be given orally.

2.   More effective against Gram-negative and less against Gram-positive than first generation cephalosporins; some are good against Bacteroides

Third generation

1.   Examples are ceftazidime, cefoperazone, cefotaxime, ceftizoxime and ceftriaxone that are given parenterally, and cefpodoxime that can be given orally.

2.   More effective against Gram-negative and less against Gram-positive than second generation cephalosporins, especially Enterobacter; some are good against Pseudomonas

3.   Moxalactam can cause extensive hemorrhage.

Fourth generation

1.   The only example available is cefepime, given parenterally.

2.   Wider range than third generation and more stable against b-lactamase

CARBAPENEMS

Imipenem

1.   A b-lactam antibiotic with a wide spectrum of activity; mode of action like penicillins and cephalosporins

2.   Used when other antibiotics fail to eradicate bacteria

3.   Given intravenously together with cilastin to prevent breakdown in the renal tubules

Aztreonam

1.   Also wide spectrum, and not usually first line antibiotic

2.   Activity mostly against Gram-negative organisms – especially Enterobacter and Pseudomonas

SULFONAMIDES AND TRIMETHOPRIM-SULFAMETHOXAZOLE (=CO-TRIMOXAZOLE)

1.   Bacteriostatic against a wide range of Gram-positive and Gram-negative, but many organisms are resistant

2.   Act by preventing synthesis of folic acid

3.   Trimethoprim and sulfamethoxazole (a sulfonamide) together are synergistic in inhibiting folic acid and bacteria show less resistance.

4.   Good oral absorption

5.   Important for patient to be well hydrated and dose reduced (or avoided) in patient with renal insufficiency

6.   Uses – used less these days, as bacteria become resistant

a.   Sulfonamides are effective against Nocardia and Toxoplasma.

b.   Can replace penicillin for prophylaxis in penicillin-sensitive patients

7.   Side effects

a.   Lack of appetite, nausea and vomiting

b.   Can cause kernicterus in neonate if given to neonate or pregnant woman

c.   Headache, depression, hallucinations

8.   Toxic effects – hypersensitivity reactions, agranulocytosis, aplastic anemia

9.   Contraindication – patients with G6PD deficiency – may cause acute hemolysis

10.   Interactions – can increase effects of oral anticoagulants, sulfonylureas and hydantoin

11.   Examples

a.   Sulfisoxazole – used topically in the eye, with erythromycin for otitis media and for urinary tract infections.

b.   Sulfasalazine is used in inflammatory bowel disease. Side-effects include nausea, fever, joint pain and rash. Toxic reactions include anemia and agranulocytosis.

c.   Sulfacetamide is used topically in the eye.

d.   Silver sulfadiazine is used on burns or decubitus (pressure) ulcers to prevent secondary bacterial or fungal infection

e.   Trimethoprim-sulfamethoxazole is used in pediatric infections and in urinary tract infections and gastroenteritis. It is also effective against Pneumocystis carinii in AIDS patients. It is used for prophylaxis in neutropenic patients.

QUINOLONES

1.   Examples are ciprofloxacin, ofloxacin and norfloxacin.

2.   Act by inhibiting DNA gyrase

3.   Well absorbed orally; also given parenterally

4.   Effective against E. coli, Shigella, Salmonella, Enterobacter, Campylobacter and Neisseria. Ciprofloxacin is more effective against

Pseudomonas, Enterococcus and Pneumococcus. Quinolones also effective against intracellular bacteria such as Chlamydia, Mycoplasma, Mycobacterium, Legionella and Brucella.

5.   Uses - urinary tract infections, prostatitis, sexually-transmitted diseases, gastroenteritis

6.   Side effects

a.   Nausea, headache, dizziness, abdominal discomfort

b.   Can increase risk of convulsions in patient with epilepsy

7.   Interactions

a.   Theophylline or non-steroidal anti-inflammatory drugs (NSAIDs) with a quinolone can cause hallucinations, delirium and even seizures.

b.   Ciprofloxacin increases concentration of theophylline.

8.   Contraindicated in children, pregnant or nursing women, and in epileptics.

DRUGS SPECIFIC FOR URINARY TRACT INFECTIONS

1.   Concentration in urine with minimal systemic effects after oral administration

2.   Examples are nalidixic acid, nitrofurantoin and methenamine.

AMINOGLYCOSIDES

General information

1.   Examples are gentamicin, netilmicin, amikacin, streptomycin, tobramycin and neomycin.

2.   Effective against aerobic Gram-negative organisms; amikacin has the widest spectrum

3.   Not absorbed orally; given parenterally or topically

4.   Uses

a.   Gentamicin, tobramycin, amikacin and netilmicin are all used in pyelonephritis, hospital-acquired pneumonia, meningitis (by Gram-negative bacteria, given intrathecally), peritonitis, sepsis.

b.   Tobramycin is also used topically in the eye.

c.   Streptomycin was the prototype; due to resistance, it is now used mostly in tuberculosis, tularemia and plague

d.   Neomycin is used only topically or orally (to clean the intestines before abdominal surgery) because of its severe toxicities.

5.   Toxic effects

a.   Ototoxicity – act on both auditory and vestibular portions of cranial nerve VIII, damage is usually irreversible, main effect is hearing loss

b.   Nephrotoxicity – usually mild and reversible; the main problem is decreased clearance increases risk of ototoxicity

c.   Rare – neuromuscular blockade with apnea

d.   Streptomycin may also cause damage to the optic nerve and peripheral neuritis

4.   Interactions - Furosemide and ethacrynic acid are synergistic for hearing loss with aminoglycosides.

TETRACYCLINES

1.   Broad spectrum – bacteriostatic against Gram-positive and Gram-negative organisms, including Rickettsia, Chlamydia, Mycoplasma pneumoniae, Brucella, Hemophilus ducreyi, Legionella, Helicobacter pylori, Trepenoma pallidum, Borrelia bungdorferi and Ureaplasma

2.	Examples are tetracycline, doxycycline, minocycline and demeclocycline.

3.	Usually given orally; can be parenterally or topically to the eye

4.	Uses – for treatment of infections by the above bacteria and also for the treatment of acne

5.	Side effects

a.	Damages intestinal flora and may cause fungal overgrowth, cause nausea, vomiting, diarrhea epigastric pain and abdominal discomfort (better if given with food); esophagitis and ulcers have been seen.

b.	Severe thrombophlebitis may occur with intravenous use.

c.	Demeclocycline and doxycycline may cause photosensitivity.

6.	Toxic effects

a.	Liver damage – especially in pregnant women

b.	Renal damage – may worsen existing disease

c.   Demeclocycline may cause nephrogenic diabetes insipidus (and so is used to treat the Syndrome of inappropriate Anti-diuretic Hormone (SIADH)).

d.   Pigmentation of the teeth may occur in small children if tetracyclines are given to them or to their mothers during pregnancy.

e.   Outdated tetracycline can cause a form of Fanconi's syndrome.

7.   Interactions – when given with dairy foods, antacids, iron compounds, sulcrafate or bismuth compounds, the metal will chelate and decrease the absorption of the antibiotic.

8. Contraindications – administration to children under 8 years old and pregnant or nursing women

CHLORAMPHENICOL

1.   Effective against a wide range of bacteria; mostly bacteriostatic; may be bactericidal against Hemophilus, Neisseria and Pneumococcus

2.   Given orally and parenterally

3.   Due to the risk of aplastic anemia, chloramphenicol is reserved for cases where the infection is resistant to other less toxic antibiotics.

4.   The "gray baby" syndrome is seen in neonates, mostly premature, who receive chloramphenicol. It is often fatal.

5.   Interactions

a.   Chloramphenicol increases half-lives of dicumarol, phenytoin, chlorpropamide and tolbutamide.

b.   Its half-life is decreased by rifampin or phenobarbital.

MACROLIDES

1.   Examples are erythromycin, azithromycin and clarithromycin.

2.   Usually bacteriostatic and effective against Gram-positive organisms, Chlamydia, Mycoplasma and Legionella; azithromycin and clarithromycin also effective against Hemophilus and Mycobacterium-avium-intracellulare.

3.   Given orally in capsules or with enteric coating as gastric acid breaks the drug down; also given parenterally (intramuscular injection is painful). Give on an empty stomach.

4.   Uses

a.   for treatment of infections by the above organisms

b.   for treatment of pertussis and prophylaxis of contacts

c.   used for prophylaxis in penicillin-sensitive patients, particularly in pregnant women

5.   Side effects

a.   Gastro Intestinal Tract (GIT) effects can affect compliance – abdominal pain, nausea and vomiting

b.   hypersensitivity

c.   Cholestatic hepatitis – which is reversible after stopping the drug

6.   Interactions – erythromycin increases effects of carbamazepine, steroids, cyclosporine, digoxin, theophylline, valproic acid, warfarin and more – via cytochrome P450

CLINDAMYCIN

1.   Bacteriostatic against Gram-positive bacteria, anaerobes, Toxoplasma and Pneumocystis carinii

2.   Given orally, topically and parenterally

3.   Uses

a.   for the treatment of infections caused by the above bacteria

b.   for the oral and topical treatment of acne

c.   for the oral and topical treatment of bacterial vaginosis

4.   Side effects

a.   Major cause of pseudomembranous colitis

b.   Also see skin rash and when given intravenously, thrombophlebitis may occur

VANCOMYCIN

1.   Effective against Staph., Strep., Enterococcus, Corynebacterium and Clostridium

2.   Given intravenously; orally for the treatment of pseudomembranous colitis

3.   Uses

a.   Treatment of pseudomembranous colitis

b.   Treatment of infections caused by Methicillin-Resistant Staphylococcus aureus (MRSA)

c.   Treatment of infections by above bacteria when other antibiotics fail

4.   Side effects - hypersensitivity

5.   Toxic effects

   a. Ototoxicity – may be reversible

a.   "Red man" syndrome – flushing, tachycardia and hypotension – occurs with too rapid intravenous infusion.

b.   May be kidney damage if given with nephrotoxic drugs or to patient with renal disorder

BACITRACIN

1.   Topical use only due to severe nephrotoxicity if given parenterally

2.   Effective against Gram-positive aerobes

3.   Uses – eyes, skin

POLYMYXIN B

1.   Topical use only due to severe nephrotoxicity if given parenterally

2.   Effective against Gram-negative bacteria, especially Pseudomonas

3.   Uses – ears, eyes and skin

DRUGS FOR THE TREATMENT OF DISEASES CAUSED BY MYCOBACTERIUM

DRUGS FOR THE TREATMENT OF TUBERCULOSIS

Background

1.   First line drugs are isoniazid, rifampin, ethambutol, streptomycin and pyrazinamide – they are the most efficient, have acceptable side effects and will treat most cases of tuberculosis in most people. Treatment is with 2 or 3 agents because use of one agent encourages resistance.   There is minimal cross-resistance between the various agents.

2.   Second line drugs are used in resistant cases or in patients with AIDS   and   include   ofloxacin,   ciprofloxacin,   ethionamide, aminosalicylic   acid,   cycloserine,   amikacin,   kanamycin   and capreomycin. They are only available for parenteral use. Due to

ototoxicity and nephrotoxicity, these agents are only used if the first line drugs are not effective. Only one should be given with a first line agent, but not with streptomycin.

3.   Different regimens are recommended. The basic regimen is isoniazid with rifampin and pyrazinamide for 2 months and then rifampin and isoniazid for another 4 months. Another possibility is isoniazid and rifampin for 9 months. The Centers for Disease Control (CDC) recommend four drugs together – isoniazid, rifampin, pyrazinamide and ethambutol or streptomycin.

4.   The CDC regimen is recommended for those exposed to resistant strains and those with tuberculosis in sites other than the lungs or extensive disease in the lungs.

5.   AIDS patients should receive at least the four-drug regimen, and in some cases a fifth or sixth drug may need to be added. Prophylactic treatment with isoniazid (ethambutol and pyrazinamide if exposed to resistant strains) is recommended to AIDS patients with positive Purified Protein Derivative (PPD) tests, anergy or at risk for tuberculosis.

6.   In other patients, prophylaxis is recommended for those exposed to tuberculosis, those with positive PPD but no clinical disease, those with history of active tuberculosis in the past who did not receive adequate treatment, and patients with anergy who come from a high-risk population. The recommended drug is isoniazid. Pregnant women and patients with active hepatic disease should receive prophylaxis at a later date.

7.   The regimen of isoniazid, rifampin and ethambutol is considered to be safe for pregnant women.

Isoniazid

1.   Drug of choice in tuberculosis

2.   Mechanism of action is unknown.

3.   Good oral and parenteral absorption; inactivation is via acetylation.   Different populations have different proportions of "rapid" and "slow" acetylators, which influence how quickly the drug is inactivated.

4.   Uses

a.   Treatment of tuberculosis; usually two or three agents together

b.   Alone for prophylaxis of tuberculosis

5.   Side effects

a.   Side effects include rash, fever, jaundice, hypersensitivity, arthritis, vasculitis, anemia, thrombocytopenia, eosinophilia and agranulocytosis.

b.   Neurological side effects include peripheral neuritis, increased risk of convulsions in epileptics, euphoria, memory loss, psychosis, muscle twitches, dizziness, ataxia, paresthesias, stupor and encephalopathy.

c.   To minimize neurological side effects, especially in patients with nutritional deficiencies (malnutrition, pregnancy, diabetes, chronic renal failure, elderly), pyridoxine (vitamin B6) is given together with the isoniazid.

d.   Hepatotoxicity, which may be fatal, is not uncommon and the risk increases with increasing age. Need to watch very carefully for signs of hepatitis, besides jaundice that is a common side effect.

6.   Toxic effects – coma, seizures, metabolic acidosis, hyperglycemia

7.   Interactions – isoniazid increases concentrations of phenytoin, especially in slow acetylators. Adjustments as necessary should be made to the dose of phenytoin without changing the dose of isoniazid.

8.   Contraindication – relative – hepatic disease

Rifampin

1.   Effective against Gram-positive bacteria such as S. aureus, Hemophilus influenzae and Neisseria meningitidis and Gram-negative bacteria such as E. coli, Proteus, Klebsiella and Pseudomonas, as well as Mycobacterium, especially M. tuberculosis, M. kansasii and M. fortuitum

2.   Acts by inhibiting RNA polymerase

3.   Uses

a.   in treatment of tuberculosis, together with isoniazid and others

b.   for prophylaxis in people exposed to meningococcemia or H. influenzae type B meningitis

c.   in cases of endocarditis or osteomyelitis resistant to other antibiotics

d.   in leprosy

4.   Given orally

5.   Side effects

a.   Colors body secretions orange

b.   Rash, fever, nausea and vomiting, jaundice

c.   Increased side effects are seen if rifampin is given infrequently or in high daily doses — a flu-like syndrome results that may develop into interstitial nephritis, acute tubular necrosis, hemolytic anemia and shock

d.   Rifampin induces hepatic microsomal enzymes and so decreases the half-lives of many, many drugs. The most important of these are digitoxin, quinidine, propanolol, oral anticoagulants, theophylline, barbiturates, oral contraceptives, halothane, methadone and sulfonylureas.

6.   Toxic effects - Hepatotoxicity, especially in children, alcoholics, the elderly and patients with liver disease

7.   Interactions

a.   Is synergistic in vitro with streptomycin and isoniazid

b.   Aminosalicylic acid may delay absorption of rifampin – best to give them 8-12 hours apart

8. Contraindications – best not to use in pregnancy or in patients with liver disease

Ethambutol

1.   Mechanism of action is not known.

2.   Oral absorption is good.

3.   Uses – in treatment of tuberculosis together with other drugs

4.   Side effects are minimal – blurry vision, rash, fever, headache, pruritus, arthralgia, dizziness, confusion, nausea and vomiting, hyperuricemia

5.   Toxic effects - optic neuritis and red-green color blindness that are dose-dependent; should test visual acuity and color blindness before and during treatment with ethambutol

6.   Contraindications – children under 5 years old

Streptomycin

1.   Aminoglycoside antibiotic

2.   Due to resistant bacteria and high toxicity, streptomycin is not often used today.

3.   It is used together with two other drugs in disseminated tuberculosis or tuberculitic meningitis.

4.   Side effects – rash, fever

5.   Toxic effects – damage to cranial nerve VIII and the kidneys

Pyrazinamide

1.   good oral absorption

2.   uses – together with other drugs for the treatment of tuberculosis

3.   side effects – hyperuricemia with clinical gout, arthralgia, nausea and vomiting, loss of appetitie, fever and painful urination

4.   Toxic effects - liver disease is common and may be fatal; need to monitor liver function regularly

5.   Contraindications – absolute – patient with liver disease

DRUGS FOR THE TREATMENT OF MYCOBACTERIUM AVIUM COMPLEX (MAC)

Background

1.   MAC infection is common in AIDS patient where it is a disseminated disease. In patients without AIDS, it is usually restricted to the lungs.

2.   Like tuberculosis, the use of one drug increases resistance and so the regiments are 2, 3 or 4 drugs.

3.   The basic regimen is ethambutol with either clarithromycin or azithromycin.

4.   Prophylaxis with rifabutin for life is recommended for AIDS patients with CD4 count under 100 cells/mm3.

Rifabutin

1.   Derivative of rifampin

2.   Used especially in AIDS patients

3. Given orally

4.   Side effects – rash, nausea and vomiting, neutropenia, orange-colored body fluids

5. Toxic effects – uveitis, arthralgia

6.   Like rifampin, induces hepatic microsomal enzymes and decreases concentrations of many drugs

Macrolides

1. Discussed in chapter on antibiotics

2. Examples are clarithromycin and azithromycin.

3.   Used in combination with another agent (ethambutol, rifampin, rifabutin, ciprofloxacin, amikacin or clofazimine)

4. Toxic effects – tinnitus, dizziness and transient loss of hearing

Quinolones

1. Discussed in chapter on antibiotics

2. Used in AIDS patients – regimen is ciprofloxacin, clarithromycin, amikacin or rifampin, ethambutol, clofazimine and ciprofloxacin

DRUGS FOR THE TREATMENT OF LEPROSY

Sulfones

1. Most widely used is dapsone; sulfoxone is a derivative of dapsone

2. Dapsone is bacteriostatic for Mycobacterium leprae and is given orally.

3. Side effects – nausea, vomiting, lack of appetite, syndrome similar to infectious mononucleosis with possible mortality

4. Toxic effects – hemolysis, methemoglobinemia

5. A sulfone syndrome – similar to the Jarisch-Herxheimer reaction – may occur, with fever, jaundice, peeling skin, anemia and swollen glands.

Clofazimine

1. Used in MAC, but mainly in leprosy, together with other agents, for dapsone-resistant strains

2. Given orally

3. One side effect seen is red coloring of the skin

DRUGS AFFECTING THE COAGULATION SYSTEM

Background

Hemostasis is the cessation of blood loss from the damaged vessel. First platelets adhere to the injured regions of blood vessel, they aggregate to form of primary hemostatic plug. Platelets stimulate local activation of plasma coagulation factors, leading to generation of fibrin clot that reinforces the platelet aggregate. Thrombosis is a pathological process in which platelets aggregate and/or fibrin clot occludes a blood vessel. Arterial thrombosis may result in ischemic necrosis of the tissue supplied by the artery. Venous thrombosis may cause tissue drained by the vein to become edematous and inflamed. Thrombosis of a deep vein may be complicated by pulmonary embolism.

Platelet aggregation and coagulation normally do not occur within an intact blood vessel. Thrombosis is prevented by several regulatory mechanisms that require a normal vascular endothelium. Prostacyclin (PGI-2), a metabolite of arachidonic acid, is synthesized by endothelial cells, and inhibits platelets aggregation and secretion. Antithrombin is a plasma protein that inhibits coagulation factors. Heparan sulfate synthesized by endothelial cells stimulate the activity of antitrombin. Protein C in combination with protein S degrade coagulating cofactors Va and VIIIa and diminishes the rates of activation of prothrombin and factor X.

ANTICOAGULANTS

1 Parenteral anticoagulants

Heparin – naturally occur in the mast cell. After releasing ingested rapidly and destroyed by macrophages.

Mode of action –

Increases rate of thrombin – antithrombin reaction, causes inhibition of factors IXa and Xa.

Effects –

High doses of heparin can interfere with platelets aggregation, whereby prolong the bleeding time and PTT.

Absorption and administration –

Not absorbed from GIT – available only with parenteral use. Administration is by continuous IV infusion, intermittent IV infusion or deep subcutaneous injection. Intravenous administration has immediate onset, subcutaneous – onset of action in 1-2 hours after injection.

There are two types of heparin in use – high-molecular-weight (HMW) and low-molecular-weight (LMW). HMW usually administrated intravenously and subcutaneously, LMW – subcutaneously.

Use –

HMW heparin in use in treatment of different types of a thromboembolism as direct anticoagulant. Useful in acute thrombotic states and prophylactically for prevention of deep vein thrombosis and thromboembolism in susceptible patients (postoperatively). Subcutaneous administration of heparin can be used for long-term treatment of patients in whom warfarin is contraindicated (e.g. during pregnancy). In these cases total daily dose of 35 000 un administered as divided doses every 8-12 hours to achieve a Partial Thromboplastin Time (test) (PTT) of 1.5 times the control values.

LMW heparin given subcutaneously only for prophylaxis of thromboembolism postoperatively or in patients with stroke. Since this has a minimal effects of clotting in vitro, routine monitoring of PTT is not needed.

Side effects and toxicity –

The primary unwanted effect is bleeding; also seen reversible thrombocytopenia, abnormalities of hepatic functions in intravenous administration, osteoporosis and allergic reactions are rare. If the presence of toxic reactions and severe bleeding persist, the heparin antagonist Protamine sulfate, is given intravenously.

2 Oral anticoagulants

Warfarin (Coumadine)

Mode of action –

Antagonists of vitamin K (activator of coagulating factors II, VII, IX and X).

Absorption and administration –

Absorption is nearly complete when administrated orally, intravenously, intramuscularly or rectally. Almost completely (99%) bound to plasma proteins, rapidly distributes in the body compounds.

The usual adult dose is 5-10 mg/d during first 2-4 days, then 2-10 mg/d. Monitoring of effectiveness is by PT values.

Side effects and toxicity –

Major effect is bleeding (especially serious is intracranial, pericardial, spinal cord bleeding, or massive internal blood loss). More rare complications are skin necrosis, alopecia (hair loss), urticaria (a skin condition, common known as hives) , dermatitis, fever, nausea, diarrhea, abdominal cramps and anorexia. In case of coumadin-induced bleeding, the vitamin K-1 analogs (phytonadione) should be administrated as antidote.

Contraindications –

Warfarin is contraindicated in pregnant women – cause fetal defects and abortions. Warfarin fetal syndrome is characterized by nasal hypoplasia, epiphyseal calcifications (intake during first trimester), CNS abnormalities (intake during second and third trimesters), fetal and neonate hemorrhage.

Interactions –

Decrease the Warfarin levels and effectiveness – barbiturates, rifampin, phenitoin, alcohol (activation of P-450), high ingestion of food, containing vit. K.

Increase the Warfarin levels and effectiveness – lowered levels of plasma proteins, vitamin K deficiency, aspirin, phenylbutazone, sulfinpyrazone, metronidazole, disulfiram, allopurinol, cimetidine, amiodarone, acute intake of ethanol.

THROMBOLYTIC DRUGS

Background

The fibrinolytic system dissolved intravascular clots as a result of plasmin, an enzyme that digests fibrin. Plasminogen, an inactive precursor, is converted in plasmin. Plasmin is a relatively nonspecific protease – it digests fibrin clots and other plasma proteins, including some coagulating factors. Tissue plasminogen activator (t-PA) is released from endothelial cells in response of various signals, including stasis produced by vascular occlusion.

Therapy with thrombolytic drugs tends to dissolve both pathological thrombi and fibrin deposits in sites of vascular injury. Therefore, the drugs are toxic, producing hemorrhage as major side effects.

Streptokinase (Kabikinase) and Streptokinase-Plasminogen complex (Antistreplase) – Streptokinase is a protein produced by b- hemolytic streptococci. It has no intrinsic enzymatic activity but it forms a stable complex with plasminogen and activate it.

Dosage and administration – a loading dose (250.000 un – 2.5 mg) must be given intravenously.

The half-life of streptokinase is 40-80 min.

Use – acute thrombotic states (mainly – coronary arteries thrombosis, acute Myocardial Infarction (MI) (heart attack))

Side effects – besides bleeding problem – allergy, rarely fever, anaphylaxis.

Urokinase (Abbokinase) – isolated from human kidney cells.

Dosage and administration – recommended administration including loading dose 1000-4500 un/kg, followed by maintenance continuous infusion of 4400 un/kg/h

The half-life of urokinase is 15-20 min, metabolized by liver.

Use – acute thrombotic states (mainly – coronary arteries thrombosis, acute MI)

Side effects – hemorrhage, allergy

Tissue Plasminogen Activator (t-PA, Alteplase) –

Binds to fibrin and activate plasminogen

The half – life is 5-10 min, metabolized in liver

Dosage and administration – given intravenously -15 mg bolus, followed by 0.75 mg/kg during next 30 min and then 0.5 mg/kg for the next hour.

Use – acute myocardial infarction

Side effects – hemorrhage

General hemorrhagic toxicity of thrombolytic drugs –

Hemorrhage in thrombolytic therapy results from two factors:

1 – the lysis of fibrin in physiological thrombi

2 – a systemic lytic state that result from systemic formation of plasmin

In case of severe hemorrhage during thrombolytic drugs administration the coagulants should be given immediately.

Contraindications for thrombolytic therapy –

-      surgery within 10 days (including trauma and CPR)

- serious GIT bleeding within 3 months

- history of hypertension (diastolic pressure higher than 110 mmHg)

- active bleeding or hemorrhagic disorder

- previous hemorrhagic CVA (stroke)

- aortic dissection

- acute pericarditis

- acute lung tuberculosis

ANTIPLATELET DRUGS

Background

Platelets provide the initial hemostatic plug at sites of vascular injury. They also participate in reactions that lead to atherosclerosis and pathological thrombosis. Antagonists of platelet function thus have been used in attempt to prevent thrombosis and to alter the natural progress of athrosclerotic vascular disease.

Aspirin – blocks production of thromboxane A-2 by covalent acetilating of cyclooxygenase (COX). The action of aspirin on platelets is permanent, lasting for the life of the platelet (7-10 days).

Dosage and administration – complete inactivation of platelet cyclooxygenase is achieved then 160 mg of aspirin taken daily. Therefor, aspirin is maximally effective as an antithrombotic drug at doses much lower than for its anti-inflammatory action. Usual dose of aspirin as antithrombotic is 160-320 mg/day. Administrated orally, mainly in form of covered tablet.

Use – treatment of acute MI (1 tab. of aspirin is given to patient for chewing), chronic treatment and prevention of thrombotic states.

Side effects – covered in chapter "NSAID's" (usually in high doses or long-term intake).

Dipyridamole (Persantine) – vasodilator; in combination with warfarin inhibits embolization from prosthetic heart valves and in combination with aspirin reduces thrombosis in patients with thrombotic diseases. Alone has no clinical benefits in use as antithrombotic drug.

Mode of action - interferes with platelets function by increasing the cellular concentrations of c-AMP.

Use – primary prophylaxis of thromboembolism in combinations with aspirin or warfarin.

Ticlopidine (Ticlid) – inhibits platelets function by producing a thrombastenia-like state.

Prolong bleeding time with maximal effect after several days of intake, abnormal platelets function persist for several days after the treatment is discontinued.

Administrated orally, usually in patient with intolerance or contraindications to aspirin.

Use – prevention of thrombosis in cerebral vascular and coronary arteries diseases.

Side effects – bleeding, nausea, diarrhea, much rarely – severe neutropenia (seen in 1% of patients).

COAGULATING DRUGS

Aminocapronic acid (Amicar) - analog of lysin that binds to plasminogen and plasmin, thus blocking the binding of plasmin to fibrin. It is a potent inhibitor of fibrinolysis and can reverse states that are associated with excessive fibrinolysis and massive hemorrhage.

Absorption and administration – rapid oral and parenteral absorption and distribution, 50% excreted with urine unchanged during 12 hours after intake. Administrated intravenously and orally; loading dose 4-5 gr given during 1 hour, followed by infusion of 1 gr/h until bleeding is controlled. No more than 30 gr should be given in 24 hours.

Use – acute and chronic hemorrhage, thrombolytic drugs overdose.

Side effects – thromboses, rarely – myopathy, muscle necrosis, allergy, hypotension, diarrhea.

Cryoprecipitates (Autoplex, Feiba) – contain coagulating factors IX, X, VII, VIII, V, and prothrombin.

Administrated intravenously.

Use – treatment of various coagulopathy states – hemophilia A and B (Christmas disease), von Willebrandt disease, etc.

Side effects – allergy, thromboses.

 **DRUGS USED IN TREATMENT OF ALLERGIES**

Antihistamines (H1 blockers)

1.   The first generation includes the

a.   ethanolamines (dimenhydrinate, diphenhydramine hydrochloride, carbinoxamine maleate and clemastine fumarate)

b.   ethylendiamines (pyrilamine maleate, tripelennamine hydrochloride, tripelennamine citrate)

c.   alkylamines (chlorpheniramine maleate, brompheniramine maleate)

d.   piperazines (hydroxyzine hydrochloride, hydroxyzine pamoate, cyclizine hydrochloride, cyclizine lactate, neclizine hydrochloride)

e.   phenothiazines (promethazine hydrochloride)

2.   The drugs of the second generation are less sedating because they do not cross the blood-brain barrier.

a.   alkylamines (acrivastine)

b.   piperazines (cetrizine hydrochloride)

c.   piperidines (astemizole, loratadine, terfenadine, levocabastine hydrochloride)

3.   Good oral absorption

4.   Effects

a.   Relaxes smooth muscle

b.   Suppresses "wheal", "flare" and itching of allergic response

c.   At high doses, has local anesthetic properties

5.   Uses

a.   To treat allergic conditions – especially allergic rhinitis, urticaria and allergic conjunctivitis

b.   Phenothiazines are used as antiemetics.

c.   Some (dimenhydrinate, diphenhydramine and promethazine) are useful in motion sickness.

6. Side effects

a. Sedation, sleepiness, decreased alertness and reactions are common. Ethanolamines are the most likely to be sedating.

b. Also seen are dizziness, tinnitus, lack of coordination, tiredness, blurry vision , nervousness, insomnia and tremors. GIT symptoms include nausea, vomiting, changes in bowel movements and lack of appetite (less so if given with food).

c. First generation agents have some anticholinergic activity and so see dry mouth, urinary retention and decreased respiratory secretions.

d. Topical use can cause local effects of hypersensitivity

7. Toxic effects

a. CNS stimulation – hallucinations, ataxia, lack of coordination, athetosis and convulsions – which also causes atropine-like symptoms (tachycardia, dry mouth, urinary retention, fever, non-responsive dilated pupils.

b. If not treated, acute toxicity can cause cardiorespiratory collapse, coma and death.

c. Large doses of terfenadine or astemizole can cause prolonged QT interval and torsade de pointes ventricular arrhythmia, especially in patients with liver disease which inhibits breakdown of the drugs.

8.   Interactions

a.   intensification of sedation with alcohol or CNS depressants

b.   Drugs that inhibit metabolism of terfenadine (macrolides, ketoconazole and itraconazole) increase risk of torsade de pointes.

DRUGS USED TO TREAT ASTHMA AND COPD

Background

1.   Asthma is due to an inflammatory process which manifests in hyperreactivity of the airways and bronchospasm. Today, therapy is based on blocking the inflammatory reaction and less on symptomatic treatment such as bronchodilators. Not all asthma is due to allergens, but there is an allergic component to asthma.

2.   Many drugs used in asthma can be delivered as aerosols, which limits their systemic effects and toxicities, while ensuring a therapeutic dose gets to the lungs, where it is needed. However, compliance is a problem and many people, particularly young children, have trouble using metered dose inhalers properly, A spacer – a device which acts as a reservoir – is useful in many of these cases. The inhaler is discharged into the spacer that in the patient's mouth, and the patient breathes as usual. If the patient still has trouble, it is best to use a nebulizer with a mask.

3.   Treatment of children with asthma is based more on cromolyn sodium than on steroids, because of the long-term effects on growth and development.

4.   Treatment of acute asthma attacks is based on b-adrenergic agonists.   Intravenous hydrocortisone or aminophylline may be necessary if the patient does not respond quickly enough.

5.   Pregnant and nursing women can be treated as any other asthmatic. Treatment is important as asthma can affect both mother and fetus. However, if possible, it is best to avoid treatment with theophylline as it can cause fetal tachycardia.

6.   Allergic rhinitis, either seasonal or perennial, is treated in a similar fashion to asthma. Part of the arsenal in treating COPD should include drugs used for the treatment of asthma, since the inflammatory component is also prominent in COPD.

GLUCOCORTICOIDS

1.   Administered orally, intramuscularly, intravenously and as aerosol for metered dose inhalers

2.   Examples used in asthma are beclomethasone dipropionate, triamcinolone acetonide and flunisolide. Budenoside dipropionate and fluticasone propionate are used to treat asthma in other countries, but are approved only for use in allergic and seasonal rhinitis in the U.S.

3.   Uses – Most cases of asthma can be maintained with aerosolized glucocorticoids, but acute attacks or severe asthma may require systemic administration.

4.   Side effects – increased risk of Candida albicans infection

5.   Toxic effects

a.   Suppression of the hypothalamic-pituitary-adrenal gland axis and so decreased production of steroids

b.   Bone loss due to resorption

c.   Thinning of skin, purpura

d.   Use of systemic glucocorticoids for 5-10 days may cause changes of mood, increased appetite, candidiasis and difficulties in balancing sugar levels in diabetics. With longer use, hypertension, abdominal striae, osteoporosis, cataracts, and psychiatric disturbances may occur.

6.   Remember that withdrawal from steroids must be gradual in order to allow the hypothalamic-pituitary-adrenal gland axis to recover.

CROMOLYN SODIUM AND NEDOCROMIL SODIUM

1.   Act by stabilizing mast cells; has no effect on bronchodilatation

2.   Available as aerosol for nebulizers and metered dose inhalers

3.   Used in the treatment of mild and moderate asthma and also in systemic mastocytosis

4.   Side effects

a.	Bronchospasm, cough, laryngeal wheezing

b.	Headache, rash, nausea, arthritis

c.	Angioedema

BRONCHODILATORS

b-Adrenergic agonists

1.	Best for symptomatic treatment of acute asthma; not as good for prophylaxis

2.	Examples

a.	Albuterol, terbutaline, pirbuterol and bitolterol are b2-selective agonists – less effects on the heart and more on the respiratory tract.

b.	Metaproterenol and isoetharine are less selective for b2.

c.	Isoproterenol is not selective.

3.   Mostly used as aerosols in metered dose inhalers; albuterol and terbutaline are also administered in nebulizers.   Oral administration is possible, but not in wide use due to increased side effects. Onset of action is quick.

4.   Side effects – mostly from oral use – muscle cramps, tachyarrhythmias, hypertension, palpitations, flushing, headache

Anticholinergic drugs

1.   Used in asthma – ipatroprium bromide

2.   Slower to act than b agonists but effect is longer lasting; when used together, there is synergism and increased effect with increased duration, and this is very effective for the management of acute asthma attacks.

3.   Not given orally

4.   Toxic effect – respiratory muscle paralysis

METHYLXANTHINES

1.   Once a mainstay of asthma therapy, the methylxanthines (theophylline, aminophylline, caffeine, and theobromine) have declined in use due to their narrow therapeutic window.

2.   A derivative of theobromine, pentoxifylline, is used in peripheral arterial disease.

3.   Good absorption with oral, rectal or parenteral use; are sustained-release tablets that have varied rates of absorption in different individuals and need to be calibrated on an individual basis.

4.   Effects

a.   Relaxation of smooth muscle, especially of the bronchi

b.   CNS stimulation – decreased fatigue and sleepiness at first; with increasing stimulation get agitation, nervousness, insomnia and hyperesthesia, progressing eventually to convulsions. There is also stimulation of respiratory and emetic centers.

c.   Diuresis

d.   Cardiovascular – increased pulse and blood pressure, premature ventricular beats may appear, but it is rare for a serious arrhythmia to result.

5.   Uses

a.   Asthma and COPD – relaxation of smooth muscle causes bronchodilatation

b.   Treatment of apnea in premature neonates

c.   Caffeine is used in preparations for the treatment of migraine and as an analgesic. The benefit is not clear.

6.   Side effects – trouble sleeping, restlessness, excitement, nausea, convulsions, tachycardia, premature ventricular beats, tachypnea

7.   Toxic effects

a.   Sudden death is not uncommon due to too rapid intravenous administration of aminophylline (give over 20-40 minutes).

b.   Less serious effects include headache, palpitations, nausea, fall in blood pressure and chest pain.

c.   Convulsions seem to be more likely with chronic intoxication. Prophylaxis with diazepam, phenytoin or phenobarbital may decrease the risk. These convulsions often are refractory and need treatment as for status epilepticus.

8.   Interactions

a.   Increased clearance of theophylline with phenytoin, barbiturates, rifampin, oral contraceptives and tobacco use

b.   Decreased clearance with cimetidine, erythromycin, interferons and during acute viral illnesses

c.   Half-life of theophylline is increased in patients with liver disease, pulmonary edema or congestive heart failure

TREATMENT OF CENTRAL NERVOUS SYSTEM DEGENERATIVE DISORDERS

The neurodegenerative diseases include common and debilitating disorders such as Parkinson's disease, Alzheimer's disease, Huntington's disease and amyotrophic lateral sclerosis (ALS). All of them are characterized by progressive and irreversible loss of neurons from specific regions of the brain. Parkinson's disease (PD) and Huntington's disease (HD) are characterized by loss of neurons from structures of the basal ganglia results in abnormalities in the control of movement. Alzheimer's disease (AD) is characterized by loss of hippocampal and cortical neurons, its lead to impairment of memory and cognitive ability. In ALS muscular weakness result from the degeneration of spinal, bulbar, and cortical motor neurons.

PARKINSON'S DISEASE (PD)

(Paralysis Agitans; Shaking Palsy)

An idiopathic, slowly progressive, degenerative CNS disorder with 4 characteristic features: slowness and poverty of movement, muscular rigidity, resting tremor, and postural instability. Parkinson's disease is the fourth most common neurodegenerative disease of the elderly. It affects about 1% of the population >= 65 yr old and 0.4% of the population >40 yr old. The mean age of onset is about 57 yr. Onset in childhood or adolescence (juvenile parkinsonism) also occurs.

Etiology and Pathophysiology

Primary parkinsonism: There is loss of the pigmented neurons of the substantia nigra, locus ceruleus, and other brainstem dopaminergic cell groups. The loss of substantia nigra neurons, which project to the caudate nucleus and putamen, results in depletion of the neurotransmitter dopamine in these areas. In postencephalitic patients the region of the midbrain containing the substantia nigra is destroyed by an inflammatory process. Onset generally is after age 40, with increasing incidence in older age groups.

Secondary parkinsonism: results from the loss of or interference with the action of dopamine within the basal ganglia, due to other idiopathic degenerative disease, drugs, or exogenous toxins. The most common cause of secondary parkinsonism is ingestion of neuroleptic drugs or reserpine. All such drugs produce parkinsonism through their dopamine-receptor-blocking properties. However, thioridazine, which has potent anticholinergic activity, is less likely to produce parkinsonism than the other traditional neuroleptic drugs. Clozapine, an atypical neuroleptic, has yet to be shown to produce parkinsonism. However, there is a low but definite incidence of agranulocytosis with this drug. Neuroleptics with the least anticholinergic activity (haloperidol) produce the greatest incidence of parkinsonism. Coadministration of an anticholinergic drug (benztropine 0.2 to 2 mg tid) or amantadine (100 mg bid) may ameliorate the condition.

Symptoms and Signs

In 50 to 80% of patients with PD, the disease begins insidiously with a resting "pill-rolling" tremor of one hand. The tremor is maximal at rest, diminishes during movement, and is absent during sleep; it is enhanced by emotional tension or fatigue. The hands, arms, and legs usually are most affected, in that order. Jaw, tongue, forehead, and eyelids may be involved as well, although the voice is not. Many patients display only rigidity and never manifest tremor. Progressive rigidity, slowness of movement (bradykinesia), and difficulty in initiating movement (akinesia) follow. Rigidity and hypokinesia may contribute to muscular aches and sensations of fatigue. The face becomes masklike and open-mouthed, with diminished blinking. The posture becomes stooped. Patients find it difficult to start walking; the gait becomes shuffling with short steps, and the arms are held flexed to the waist and fail to swing with the stride. The steps may inadvertently quicken and the patient may break into a run to keep from falling. A tendency to fall forward (propulsion) or backward (retropulsion) when the center of gravity is displaced results from the loss of postural reflexes. Speech becomes hypophonic, with a characteristic monotonous, stuttering dysarthria. Hypokinesia and impaired control of distal musculature results in micrographia and increasing difficulties with activities of daily living.

The sensory examination usually is normal. Signs of autonomic nervous system dysfunction (eg, seborrhea, constipation, urinary hesitancy, and orthostatic hypotension) may be seen. Dementia occurs in about 50% of patients; depression also is common. In some patients dementia and depression constitute major aspects of the disability.

Treatment

Levodopa - the metabolic precursor of dopamine, crosses the blood-brain barrier into the basal ganglia where it is decarboxylated to form dopamine, replacing the missing neurotransmitter. Bradykinesia and rigidity are the symptoms helped most, although tremor is often substantially reduced. Mildly affected patients may return to nearly normal, and bedridden patients may become ambulatory. Extensive peripheral metabolism of the drug has 2 consequences; it must be given in very large doses, and side effects (nausea, palpitations, flushing) may be severe. Coadministration of the peripheral decarboxylase inhibitor carbidopa lowers dosage requirements by preventing catabolism, thus decreasing side effects and allowing more efficient delivery of levodopa to the brain. Carbidopa/levodopa is available in fixed-ratio preparations of 10/100, 25/100, 25/250 mg, and a controlled-release tablet of 50/200 mg.

Treatment is begun with a single 25/100 tablet tid. The dosage is gradually increased every 4 to 7 days according to patient tolerance until maximum benefit is reached. Side effects may be minimized by gradually and carefully increasing the dosage and by giving the drug with or after meals. (However, large amounts of protein may interfere with absorption of levodopa.) Most patients require 400 to 1000 mg/day of levodopa in divided doses q 2 to 5 h. At least 100 mg/day of carbidopa is necessary to minimize peripheral side effects. Some patients may require up to 2000 mg/day of levodopa (and 200 mg of carbidopa).

Side effects

Involuntary movements (dyskinesias) in the form of oral-facial or limb chorea or dystonia are often the dose-limiting side effects of levodopa therapy. The threshold for their emergence seems to decrease with length of treatment. In some patients effective reduction of parkinsonism cannot be achieved except at the price of some degree of dyskinesia. After 2 to 5 yr of treatment, >50% of patients begin to experience fluctuations in their response to levodopa ("on-off effect"). The duration of improvement following each dose of drug shortens, and superimposition of dyskinetic movements results in swings from intense akinesia to uncontrollable hyperactivity. Such swings have traditionally been managed by keeping individual doses of levodopa as low as possible, using dosing intervals as short as 1 to 2 h. Dopamine agonist drugs, controlled-release levodopa/carbidopa, or selegiline may be useful in the treatment of this problem. Other side effects of levodopa include orthostatic hypotension, hallucinations, and occasionally toxic delirium.

Amantadine 100 to 300 mg/day is useful in treating early, mild parkinsonism in 50% of cases, and to augment the effects of levodopa later in the illness. Its mechanism of action is uncertain; it may act through augmentation of dopaminergic activity, anticholinergic effects, or both. Amantadine often loses its effectiveness after a period of months when used as a single agent.

Side effects include lower extremity edema, livedo reticularis, and con-fusion.

Bromocriptine and pergolide are ergot alkaloids that possess antiparkinsonian activity because they directly activate dopamine receptors in the basal ganglia. Bromocriptine 5 to 60 mg/day or pergolide 0.1 to 7.0 mg/day are useful at all stages of the illness. Their most traditional use comes in the later stages when response to levodopa diminishes or on-off effects are prominent. In such cases, usefulness is often limited by a high incidence of side effects including nausea, orthostatic hypotension, confusion, delirium, and frank psychosis. Such side effects may be controlled by reducing the dose of levodopa. Recently, evidence has accumulated showing that the use of bromocriptine or pergolide early in treatment, in conjunction with small doses of levodopa, may delay the emergence of drug-induced involuntary movements and on-off effects. This is perhaps due to the long half-lives of the synthetic drugs. Prolonged dopamine receptor stimulation is more physiologic than that due to levodopa, which has short plasma half-life. This results in preservation of the integrity of postsynaptic dopamine receptors and a more normal drug response. However, rarely can either bromocriptine or pergolide be used as a sole antiparkinsonian agent; concomitant administration of levodopa is almost always necessary.

Monoamine oxidase type B (MAO-B) inhibitors

Selegiline - inhibits one of the 2 major enzymes responsible for the breakdown of dopamine in the brain, thereby prolonging the action of individual doses of levodopa. At doses of 5 to 10 mg/day, it is devoid of the hypertensive crisis ("cheese effect") common to the nonselective MAO inhibitors, which block both the A and the B isoenzymes. Selegiline is useful in diminishing the end-of-dose wearing off of levodopa effect in some patients with mild on-off problems. It is virtually devoid of its own side effects, but it can potentiate the dyskinesias, mental side effects, and nausea produced by levodopa, and the dose of levodopa may need to be reduced.

Anticholinergic drugs

Were the mainstay of antiparkinsonian treatment before dopaminergic drugs. Currently they are used alone in the early stages of treatment and later to supplement levodopa. Commonly used anticholinergics include benztropine 0.5 to 2 mg tid and trihexyphenidyl 2 to 5 mg tid. As with levodopa, initial dosage should be small and dosage should be increased as tolerated.

Adverse effects - dry mouth, urinary retention, constipation, and blurred vision. Particularly troublesome in older patients are confusion, delirium, and impaired thermoregulation due to decreased sweating.

Antihistamines with anticholinergic action

Diphenhydramine 25 to 100 mg/day and Orphenadrine 50 to 200 mg/day - are useful for treating tremor and as mild sedatives.

Tricyclic antidepressants

Amitriptyline - used in low doses such as 25 to 50 mg at bedtime) often are useful as nighttime sedatives and as adjuvants to levodopa, in addition to their effectiveness in treating depression.

Propranolol (10 mg bid to 40 mg qid) occasionally is helpful when parkinsonian tremor is accentuated rather than quieted by activity or intention.

ALZHEIMER'S DISEASE

Alzheimer-type dementia is a degenerative process, with a loss of cells from the basal forebrain, cerebral cortex, and other brain areas. Acetylcholine-transmitting neurons and their target nerve cells are particularly affected. The brain shows moderate to marked atrophy with wide sulci and dilated ventricles. Memory loss is the most prominent early symptom. Disturbances of arousal do not occur early in the course. Alzheimer's presenile and senile onset dementias are similar in both clinical and pathologic features, with the former commonly beginning in the 5th and 6th decades and the latter in the 7th and 8th decades, sometimes earlier, rarely later. The dementia usually progresses steadily, becoming well advanced in 2 to 3 yr.

Treatment

Tacrine (Cognex) – acridine derivative; potent centrally acting inhibitor of AchE, administrated orally.

Has mild clinical effects in early stages of the disease.

Side effects – dose-dependent: nausea, vomiting, abdominal cramps, diarrhea, hepatotoxicity (elevation of the liver enzymes levels).

HUNTINGTON'S DISEASE

Huntington's disease or chorea - is an autosomal dominant disorder usually beginning in middle age (35 to 50 yr) and characterized by choreiform movements and progressive intellectual deterioration. Both sexes are affected equally. Pathologically there is atrophy of the caudate nucleus with degeneration of the small-cell population and decreases in levels of the neurotransmitters gamma-aminobutyric acid (GABA) and substance P.

Symptoms and signs

Onset is insidious. Psychiatric disturbances, ranging from personality changes of apathy and irritability to full-blown manic-depressive or schizophreniform illness may precede the movement disorder or develop during its course. Anhedonia or asocial behavior may be the first behavioral manifestations of the disease. Motor manifestations include flicking movements of the extremities, a lilting gait, and motor impersistence (inability to sustain a motor act such as tongue protrusion). Facial grimacing, ataxia, and dystonia may also appear.

Prognosis and treatment

The illness is progressive; no treatment is known. Patients ultimately lose physical and mental abilities to care for themselves. The choreic movements and agitated behaviors may be suppressed, but usually only partially, by phenothiazine (chlorpromazine 100 to 900 mg/day) or butyrophenone neuroleptics (haloperidol 10 to 90 mg/day) or reserpine, beginning with 0.1 mg/day and increasing until side effects of lethargy, hypotension, or parkinsonism supervene. For treatment of depression and irritability fluoxetine and carbamazepine are useful. Therapeutic strategies to replace brain GABA stores have been ineffective.

AMYOTROPHIC LATERAL SCLEROSIS (ALS)

Motor neuron disease of unknown cause characterized by progressive degeneration of corticospinal tracts and/or anterior horn cells and/or bulbar motor nuclei. The symptoms and descriptive designation vary according to the part of the nervous system most affected. Median age of onset is age 55, and the incidence is greater in males; 5% of cases are famil-ial, with an autosomal dominant inheritance.

Symptoms and Signs

Muscular weakness and atrophy, evidence of anterior horn cell dysfunction, are most often noted initially in the hands and less often in the feet. The site of onset is random and progression is asymmetric. Cramps are common and may precede weakness. Visible muscle fasciculations, spasticity, hyperactive deep tendon reflexes, and extensor plantar reflexes, evidence of corticospinal tract involvement, soon accompany the lower motor neuron signs. Dysarthria and dysphagia are due to involvement of brainstem nuclei and pathways. Sensory systems, voluntary eye movements, and urinary sphincters are spared. A rare patient may survive for 30 yr: 50% die within 3 yr of onset, 20% live 5 yr, and 10% live 10 yr.

Treatment

There is no specific treatment. Baclofen may help reduce spasticity; quinine or phenytoin may be used to decrease cramps. Tricyclic antidepressants may be used to decrease saliva production (amitriptyline 10 mg orally qid).

DRUGS FOR THE TREATMENT OF VIRAL DISEASES

Acyclovir and valacyclovir

1.   Acts against herpes simplex viruses type 1 and 2 (HSV-1, HSV-2); is also some activity against herpes zoster virus (HSV-3).

2.   Valacyclovir is a prodrug that yields acyclovir.

3.   Mode of action – inhibition of viral DNA synthesis

4.   Can be administered orally, intravenously or as topical ointment; oral bioavailability is under 30%. Crosses the placenta and can be found concentrated in breast milk.

5.   Uses

b.   Decreases intensity, duration and frequency of herpes simplex attacks

c.   Topical preparation less effective

d.   Given intravenously as prophylaxis in carriers who receive immunosuppression, for HSV encephalitis and for varicella pneumonia and encephalitis

e.   Valacyclovir has been found useful for the pain of herpes zoster (Chicken Pox), but not post-herpetic neuralgia

6.   Side effects

One.   Acyclovir – minimal side effects – nausea, diarrhea, rash, headache

Two.   Valacyclovir – headache, nausea, diarrhea

7.   Toxic effects

One.   Acyclovir – renal failure, tremor, altered sensorium, delirium, seizures, extrapyramidal signs

Two.   Valacyclovir – kidney damage, thrombocytopenia (in immunocompromised patients)

8.   Interactions

One.   Acyclovir with zidovudine may cause deep lethargy and sleepiness.

Two.   Acyclovir with cyclosporine (or other nephrotoxic drug) has increased risk of nephrotoxicity

Three. Probenecid causes increased activity of acyclovir.

Four.   Acyclovir increases activity of methotrexate.

Famciclovir and penciclovir

1.   Famciclovir is a pro-drug of penciclovir.

2.   Same mode of action as acyclovir

3.   Penciclovir has very low oral bioavailability; famciclovir can be given orally with better results

4.   Uses – herpes zoster in immunocompromised adults

5.   Side effects – headache, diarrhea, nausea

6.   Toxic effects – may increase risk of tumor and cause mutations

7.   Interactions – famciclovir increases concentrations of digoxin

Foscarnet

1.   Acts by inhibiting synthesis of viral nucleic acids – works on DNA polymerase (of the herpes viruses) or reverse transcriptase (of the HIV virus)

2.   Oral bioavailability is low; also given intravenously

3.   Uses

One.   cyto-megalo virus (CMV) retinitis and other CMV infections

Two.   CMV resistant to ganciclovir

Three. HSV resistant to acyclovir

4.   Side effects – headache, tremor, seizures, hallucinations, irritability, fever, nausea, vomiting, leukopenia, genital ulcers and changes in liver function tests.

5.   Toxic effects – clinical hypocalcemia, acute tubular necrosis, interstitial nephritis

6.   Interactions – when given with pentamidine, risk of hypocalcemia is increased

Ganciclovir

1.   Inhibits synthesis of viral DNA

2.   Especially effective against CMV

3.   Low oral bioavailability; given orally or intravenously

4.   Uses

One.   Treatment of CMV retinitis

Two.   Prophylaxis of CMV infection in immunocompromised patients

5.   Side effects – headache, phlebitis, rash, anemia, fever, nausea, vomiting, changes in liver function tests

6.   Toxic effects – bone marrow suppression, convulsions, coma

7.   Interactions

One.   Risk of bone marrow suppression increased with concurrent use of zidovudine, nephrotoxic drugs and cytotoxic drugs

Two.   Probenecid decreases clearance of ganciclovir.

Zidovudine

1.   Inhibits reverse transcriptase by competing with thymidine; inhibits retroviruses

2.   Good oral absorption and bioavailability

3.   Uses

One.   in HIV positive patients with CD4 counts of under 500/mm3

Two.   in pregnant women with AIDS and their neonates to prevent vertical transmission

4.   Major problem with zidovudine is that there is tolerance after about a year

5.   Side effects – bad headache, nausea, vomiting, difficulties in sleeping, muscle aches

6.   Safety in pregnancy not completely clear – neonates show anemia and growth retardation, but no excess malformations found

7.   Toxic effects – granulocytopenia and anemia

8.   Interactions

One.   Increased risk of bone marrow suppression with fluconazole, ganciclovir or probenecid

Two.   Reduced consciousness may occur with acyclovir.

Three. Clarithromycin decreases absorption of zidovudine.

Four. Rifampin may decrease concentration of zidovudine.

Didanosine

1.   Acts via competitive inhibition of reverse transcriptase

2.   Given orally, regular or buffered tablet, or intravenously

3.   Uses – advanced HIV infection after at least 4 months of zidovudine

4.   Side effects – diarrhea, rash, headache, seizures, difficulty sleeping, optic neuritis, loss of retinal pigmentation in children, hyperuricemia

5.   Toxic effects – peripheral neuropathy, pancreatitis

6.   Interactions – buffered tablet can decrease absorption of ketoconazole, dapsone, tetracycline, quinolones

Stavudine

1.   Competitive inhibition of reverse transcriptase

2.   Well-absorbed orally

3.   Uses – in AIDS where other agents have not helped

4.   Side effects – anemia, arthralgia, rash, fever

5.   Toxic effects – peripheral sensory neuropathy, elevated liver enzymes

6.   Interactions - avoid giving with didanosine or zalcitabine because of increased risk of neuropathy.

Zalcitabine

1.   Competitive inhibition of reverse transcriptase

2.   High oral bioavailability; also given intravenously

3.   Uses

One.   Together with zidovudine for AIDS patients with CD4 under 300/mm3

Two.   For AIDS patients intolerant to zidovudine

4.   Side effects – fever, nausea, rash, headache, mouth ulcers, granulocytopenia

5.   Toxic effects – peripheral neuropathy

6.   Interactions – avoid use of didanosine and stavudine.

Amantadine and rimantadine

1.   Inhibit replication of influenza A viruses

2.   Good oral absorption

3.   Uses – treatment and prophylaxis of influenza A; amantadine is also used in Parkinson's disease

4.   Side effects – nausea, anorexia, nervousness, lack of concentration, difficulty sleeping

5.   Toxic effects of amantadine - delirium, hallucinations, seizures, coma, arrhythmias, worsening of psychiatric problems and epilepsy

6.   Interactions – toxic effects on the CNS worsened by use with antihistamines, neuroleptic drugs or anticholinergic drugs

Interferons

1.   Interferons are cytokines and are divided into a, b, and g interferons. One of their functions is in increasing resistance to viruses, either directly or via modulation of the immune response.

2.   Oral bioavailability is nil; intramuscular or subcutaneous administration is effective.

3.   Uses

One.   Human papilloma virus – genital warts

Two.   Chronic hepatitis B and C

Three. Multiple sclerosis

Four.   Kaposi's sarcoma

4.   Side effects – flu-like syndrome, elevated liver enzymes, interstitial nephritis, proteinuria, hair loss

5.   Toxic effects – myelosuppression, sleepiness, confusion, fatigue, weight loss, thyroid problems and cardiac damage

6.   Interactions – increases myelotoxicity of zidovudine

Ribavirin

1.   Inhibits synthesis of viral m-RNA

2.   Fair oral absorption; also given intravenously, but mostly as aerosol

3.   Uses

a.   Respiratory infections due to RSV (respiratory syncytial virus)

b.   Severe influenza

c.   Parainfluenza and measles in immunocompromised patients

4.   Side effects

One.   Aerosol preparation can cause irritation to the eyes, rash and wheezing.

Two.   Systemic side effects include anemia, elevated bilirubin, iron and uric acid.

5.   Toxic effects – rigors

6.   Ribavirin appears to be teratogenic, embryotoxic, oncogenic and gonadotoxic.

DRUGS FOR THE TREATMENT OF FUNGAL INFECTIONS

Amphotericin B

1.   Can only be given parenterally; new preparations based on liposomes are in use in other countries

2.   Effective against Candida, Cryptococcus, Blastomyces, Histoplasma, Coccidioides, Aspergillus and mucormycosis

3. Side effects

a.   Fever and chills are most common. Also seen are nausea, vomiting, headache, weight loss and phlebitis (inflammation of a vein).

b.   Respiratory distress and a slight fall in blood pressure may occur – this may be serious in patients with cardiac or respiratory disease; best to pretreat such patients with paracetamol/acetaminophen or intravenous hydrocortisone hemisuccinate.

4. Toxic effects – azotemia is very common and is dose-dependent; renal tubular acidosis and loss of magnesium and potassium in urine also occur

Flucytosine

1. Effective against Cryptococcus, Candida and chromomycosis.

2. Given orally

3. Resistance develops quickly and so should not be used as single agent.

4. Side effects – rash, nausea, vomiting, diarrhea, enterocolitis, transient elevation of liver enzymes

5. Toxic effects - myelosuppression

Ketoconazole

1. Other similar antifungals are clotrimazole, miconazole, econazole, butoconazole and sulconazole.

2. Effective against Blastomyces, Histoplasma, Coccidioides, ringworm, tinea versicolor, Candida vulvovaginitis, oral and esophageal Candida and chronic mucocutaneous candidiasis

3. Given orally; metabolized by cytochrome P450 and cleared by hepatic microsomal enzymes. Can also be given intrathecally for brain mycoses.

4. Used also to decrease androgen production in women with polycystic ovary disease and cortisol production in Cushing's disease

5. Side effects

a.   Vomiting and loss of appetite are most common. Also seen are rash and pruritus.

b.   Transient elevations in liver enzymes are common; full-blown hepatitis is rare but may be fatal

6. Toxic effects

a.   ketoconazole inhibits synthesis of steroids and so can cause menstrual irregularities, gynecomastia, lowered libido

b.   because of the risk of Addison's disease due to inhibited steroidogenesis, best not to use ketoconazole in patients with burns, trauma or before surgery

7. Interactions

a.   Bioavailability is decreased when given with H2 blockers or antacids – due to less acidic environment.

b.   Rifampin and phenytoin decrease concentration of ketoconazole.

c.   Ketoconazole increases concentrations of cyclosporine (via cytochrome P450); it also increases concentrations of terfenadine and astemizole and can cause torsades de pointes

8. Not recommended during pregnancy or lactation

Itraconazole

1. Other similar antifungals are terconazole and fluconazole.

2. Very similar to ketoconazole with a wider range of effectiveness and fewer side effects

3. Uses – drug of choice for histoplasmosis (other than meningeal); also used in candidiasis, onychomycosis, ringworm and tinea versicolor

4. Given orally

5. Side effects – nausea, vomiting, increased blood lipids, hypokalemia, rash,

6. Toxic effects – adrenal insufficiency, edema of legs, hypertension

7. Interactions

a. like ketoconazole, with rifampin, phenytoin and carbamazepine; with antacids, proton pump inhibitors and H2 blockers, and with digoxin and cyclosporine

b. Don't give with didanosine since it also reduces gastric acid.

Fluconazole

1. Effective against Candida, Cryptococcus, Coccidioides, Histoplasma, Blastomyces, sporotrichosis, ringworm

2. Excellent oral absorption; also given parenterally

3. Uses

a. candidiasis

b. cryptococcal meningitis in AIDS patients

c. coccidioidal meningitis

4. Side effects – nausea, vomiting, headache, rash, abdominal pain, diarrhea, hair loss

5. Interactions

a. Fluconazole increases concentration of phenytoin, zidovudine, rifabutin, cyclosporine, warfarin and sulfonylureas.

b. Rifampin decreases fluconazole concentration, but not significantly.

Griseofulvin

1.   Effective against Microsporum, Epidermophyton and Trichopyton (dermatophytes)

2. Given orally; since very poor aqueous solubility given as powder

3. Uses – ringworm, athlete's foot – if topical preparations have not helped

4. Side effects

a. minimal – most common is headache

b. less common are peripheral neuritis, fatigue, syncope, vertigo, blurry vision, confusion, nausea, vomiting, diarrhea, flatulence, leukopenia, neutropenia

5. Interactions – induces hepatic microsomal enzymes

Topical antifungals

1. Used for superficial fungal infections – candidiasis, ringworm, tinea versicolor, fungal keratitis, athlete's foot

2.   Not usually effective in fungal infections of the nails (onychomycosis) or hair

3.   Examples are clotrimazole, econazole, miconazole, ciclopirox olamine, haloprogin, terbinafine and Nystatin

4.   Applied to skin or inserted into the vagina; oral formulations are available for oropharyngeal candidiasis

5.   Side effects

a.   local reactions on skin and vagina

b.   Systemic effects are almost non-existent.

6.   These antifungals can be used in pregnancy.

DRUGS FOR THE TREATMENT OF HELMINTHIASIS (PARASITIC WORMS)

Benzimidazoles

1.   Examples are mebendazole, albendazole and thiabendazole.

2.   Act by inhibiting microtubule polymerization

3.   Thiabendazole has rapid oral absorption; mebendazole and albendazole are poorly absorbed.

4.	Uses – nematodes of the GIT; albendazole is not available in the USA but in other countries is the drug of choice for cysticercosis and cystic hydatid disease.

5.	Side effects of thiabendazole

One.	Most common are loss of appetite, nausea, vomiting and dizziness.

Two.	Also seen are diarrhea, fatigue, sleepiness and headache.

6.	Mebendazole has minimal side effects.

7.	Avoid use in patients with liver disease, potentially hepatotoxic.

8.	Safety not established in pregnant women or children in the first two years.

Praziquantel

1.	Acts on the neuromuscular system and activates host defenses against the worms

2.	Oral absorption is good

3.	Effective against cestodes and trematodes (not effective against nematodes).

4.   Uses in the USA – schistosomiasis and liver flukes

5.   Side effects

One.   Abdominal pain, nausea, headache, dizziness, sleepiness

Two.   May be fever, urticaria, pruritus joint and muscle pains and eisonophilia

Three. In CNS infections – inflammatory reactions may cause meningismus, seizures, cells in the CSF and changes in mental status

6.   Interactions - bioavailability is reduced by carbamazepine and phenobarbital; it is increased by cimetidine and dexamethasone.

7.   Contraindications – do not give in cases of ocular cysticercosis – may be irreversible damage.

DRUGS USED IN MALARIA

Chloroquine and hydroxychloroquine

1.   Effective against all four types of malaria when in red blood cells, but not against gametocytes of Plasmodium falciparum and not against latent forms of P. ovale or P. vivax; many strains of P. falciparum are resistant to chloroquine.

2.   Good oral absorption; also given parenterally

3.   Uses

One.   Malaria – chloroquine is the main drug for use in treatment and in prophylaxis

Two.   Hydroxychloroquine is used also for mild rheumatoid arthritis, amoebic infections of the liver and systemic lupus erythematosus.

4.   Side effects

One.   Headache, nausea and vomiting, urticaria, blurred vision, diplopia, wide QRS

Two.   Pruritus is common in persons with dark skin

5.   Toxic effects

One.   Hypotension, changes in ECG, myocardial depression and even cardiac arrest

Two.   Ototoxicity

Three. Retinopathy – at doses above 250 mg/day

Four. Myopathy, peripheral neuropathy, cardiopathy

6.   Contraindications

One.   Patients with hepatic, gastrointestinal, neurological or hematological disorders

Two.   Patients with G6PD deficiency

Three. Patients with porphyria cutanea tarda or psoriasis

7.   Interactions – increased risk of dermatitis if given with gold or phenylbutazone

Quinine

1.   Acts on forms in the red blood cells and gametes of P. vivax and P. malariae

2.   Given orally, intravenously or intramuscularly

3.   Uses

One.   Treatment of malaria due to chloroquine-resistant P. falciparum

Two.   Blocks the action of physostigmine on muscles and is used for symptomatic treatment in myotonia congenita and in nocturnal muscle cramps.

4.   Side effects

One.   High levels of insulin cause hypoglycemia.

Two.   Hypersensitivity reactions are common.

5.   Toxic effects are called cinchonism.

One.   Tinnitus, vertigo, headache, nausea and disturbances of vision (diplopia, photophobia, night blindness, scotomas, dilated pupils) are the first signs.

Two.   Later, GIT disturbances (nausea, vomiting, diarrhea), sweating, flushed face and rashes appear.

Three. Arrhythmias are also common.

Four.   "Blackwater fever", characterized by massive hemolysis with hemoglobinuria and hemoglobinemia, occurs in pregnant women treated with quinine or quinidine, causes anuria, renal failure and can be fatal. It is rare.

6.   Contraindications

One.   Patients with myasthenia gravis (can cause dysphagia and difficulties in breathing)

Two.   Patients with G6PD deficiency

Three. Patients with tinnitus or optic neuritis

7.   Interactions

One.   Compounds containing aluminum decrease oral absorption of quinine..

Two.   Quinine increases concentrations of digoxin and warfarin and enhances effects of neuromuscular blocking agents. It antagonizes acetylcholinesterase inhibitors.

Three. Cimetidine decreases renal clearance of quinine and acid urine increases it.

Primaquine

1.   Effective against hepatic and latent tissue forms of P. vivax and P. ovale

2.   Given orally because parenteral administration decreases blood pressure; oral absorption is good

3.   Uses – recurrent attacks of malaria, best given together with chloroquine

4.   Side effects – epigastric and abdominal discomfort,

5. Toxic effects – methemoglobinemia, leukopenia

6. Contraindications

One. Patients with G6PD deficiency

Two. Patients with acute illness and risk of granulocytopenia (i.e. active rheumatoid arthritis or active systemic lupus erythematosus)

Three. Use together with other drugs which affect hematopoiesis or hemolysis

Pyrimethamine

1. Effective against acute attacks of falciparum malaria

2. Slow oral absorption

3. Uses

One. Malaria – usually in a preparation also containing a sulfonamide for synergism – mostly for chloroquine-resistant strains

Two. Together with sulfadiazine is used to treat toxoplasmosis

4. Toxic effects – megaloblastic anemia (due to folate deficiency)

Halofantrine

1. Acts against P. falciparum in the red blood cells, but no effect on latent forms or gametocytes; most effective against mature forms

2. Slow oral absorption; minimally soluble in water so can't be used parenterally

3. Uses

One. Acute attack of resistant falciparum malaria

Two. Should not be used for prophylaxis since its slow elimination allows development of resistant strains

4. Side effects – nausea, vomiting, diarrhea, abdominal pain; pruritus and rash may occur in persons with dark skin

5. Toxic effects may occur even at low doses due to unpredictable absorption and bioavailability. They include arrhythmias and prolonged QT interval.

Mefloquine

1.   Effective against P. falciparum

2.   Good oral absorption; parenteral use causes serious local reactions

3.   Uses – treatment and prophylaxis of chloroquine-resistant P. falciparum

4.   Side effects

One.   Vomiting is very common.

Two.   Also see nausea, diarrhea, abdominal pain and mild dizziness

5.   Toxic effects

One.   Headache, ataxia, severe dizziness, hearing and visual disturbances are mild and self-limited.

Two.   Seizures, encephalopathy, psychotic behavior and disorientation are more serious, but usually reversible upon stopping the drug.

6.   Contraindications

One.   Pregnant (especially first trimester) and nursing women

Two.   Children less than 2 years old (or less than 15 kg body weight)

Three. Patients with neurological or psychiatric disorders

7.   Interactions

One.   Increased risk of convulsions and arrhythmias if given with quinine, quinidine or chloroquine

Two.   Increase risk of convulsions in patients treated with valproic acid

Other drugs used in the treatment of malaria include sulfonamides and the tetracyclines, which are used together with the above drugs.

DRUGS USED IN OTHER PROTOZOAL DISEASES

Metronidazole

1.   Effective against protozoans such as Trichomonas vaginalis and Entamoeba histolytica but also against anaerobic bacteria.

2.   Given orally, intravenously and topically to the vagina and skin

3.   Uses

One.　　Anaerobic bacteria including Bacteroides, Clostridium and Helicobacter

Two.　　Used in the treatment of pseudomembranous colitis

Three. Used in Trichomonas infections in both men and women

Four.　Used also in amebiasis and giardiasis

4.　　Side effects

One.　　Most common are headache, nausea, metallic taste in the mouth and dryness of the mouth.

Two.　　Vomiting and diarrhea may also occur.

Three. Neutropenia may be seen and it is reversible.

5.　　Toxic effects

One.　　Dizziness, vertigo, encephalopathy, convulsions, ataxia

Two.　　Numbness or paresthesias are rare, but may not be completely reversible.

6.　　Interactions

One.   Metronidazole's metabolism in the liver is decreased by cimetidine and increased by phenobarbital, rifampin and prednisone.

Two.   Avoid alcohol since metronidazole has a disulfiram effect. Avoid disulfiram since psychosis or confusion may occur.

7. Use in the first trimester of pregnancy is not advised.

Pentamidine

1.   Effective against trypanosomes and Pneumocystis carinii

2.   Given intramuscularly or intravenously for treatment; as aerosol for prophylaxis (less systemic toxicity)

3.   Uses

One.   Main use is for AIDS patients with Pneumocystis carinii – treatment and prophylaxis.

Two.   Together with suramin is used in treatment of West African trypanosomiasis.

Three. Also used in visceral leishmaniasis

4.   Side effects – rash, thrombophlebitis and sterile abscesses at site of injection, thrombocytopenia, anemia, neutropenia, renal insufficiency

5.   Toxic effects

One.   Can be immediate and serious – breathlessness, increased heart rate, headache, vomiting, dizziness, syncope

Two.   Pancreatitis, hypoglycemia or hyperglycemia and even insulin-dependent diabetes may occur

6. Interactions – risk of hypocalcemia if given with foscarnet

CHEMOTHERAPY OF NEOPLASTIC DISEASES

Background

In modern medicine cytotoxic drugs are broadly and successfully in use not only in treatment of neoplastic diseases but also in many other clinical situations: as component of immunosuppressive regiments for rheumatoid arthritis (methotrexate and cyclophsphmide), organ transplantation (methotrexate and azathioprine), sickle cell anemia (hydroxyurea), anti-infective chemotherapy (trimetrexate and leucovorine), and psoriasis (methotrexate).

Many of cytotoxic drugs act at specific phases of the cell cycle and have activity only against the cells that are in process of division. These group of agents called cell-cycle specific (CCS). On other hand, there is the cytotoxic agent that affect cell without connection to phase of cell cycle, they called non-cell-cycle specific (NCCS).

Cell cycle divided on four phases – 1) G1-phase – presynthetic phase; 2) S-phase – DNA synthesis (this phase affect such agents as cytarabine, hydroxyurea, 6-mercaptopurine, methotrexate); 3) G2-phase – premitotic interval; 4) M-phase – mitotic phase (this phase affect such agents as vinctistine, vinblastine, paclitaxel).

Non-cell-cycle specific agents affect neoplastic cells in any phase. This group include alkylating drugs, nitrosurea, antitumor antibiotics, procarbazine, cis-platinum, dacarbasine.

A greatest problem of neoplastic diseases management is a tumor cell resistance to chemotherapy. Most commonly such cells appears after treatment with single cytotoxic agent. The resistance may be specific to the selected agent or more general to a broad range of cytotoxic drugs. To tumor cell resistance may lead a mutation or loss of p53 suppressive oncogen (a suppressor gene is essential for normal control of cell proliferation; its loss or mutation allows cell to undergo malignant transformation). Mutation of p53 or its loss inactivate the key pathway of programmed cell death and leads to continuous proliferation of tumor cells that have the capacity to survive DNA damage.

There are many types of tumor cells resistance, depend on type of activity of the cytotoxic agent.

I.   ALKYLATING AGENTS

Mode of action – formation of covalent linkages by alkylation of various nucleonic structures (phosphates, amino, sulfhydryl, hydroxyl, carboxyl, imidazole groups), thus disturb fundamental mechanisms of cell proliferation. These compounds may alkylate nondividing cells, but cytotoxicity is markedly enhanced if DNA in cell is programmed to divide.

Mechanisms of tumor cells resistance – 1) decreased permeation of drug; 2) increased production of nucleophylic substances, that can compete with drugs for DNA; 3) increased activity of DNA repairing enzymes.

162

Nitrogen mustards

Mechlorethamide (Mustargen) – most reactive in this class

Absorption and administration – given intravenously, rapid absorption and distribution

Therapeutic use – mainly in treatment of Hodgkin's disease

Clinical toxicity – major manifestations are nausea, vomiting, lacrimation, myelosuppression. Also seen reproductive system disturbances (menopause, oligospermia), teratogenic (not in use in first trimester of pregnancy); local manifestations – thrombophlebitis, in extravasation during infusion – severe in duration.

Cyclophosphamide (Cytoxan)

Absorption and administration – good oral absorption; activated by P-450; rapid distribution in oral and parenteral (intravenous) intake.

Therapeutic use – clinical spectrum of activity is very broad (non-Hodgkin's lymphomas, breast cancer, multiple myeloma, carcinomas of lung, cervix, ovary, neuroblastoma, etc.)

Clinical toxicity - nausea, vomiting, mucosal ulcerations, hemorrhagic cystitis, increased skin pigmentation; less common – lung fibrosis; rare. No local manifestations in drug extravasation.

Ifosfamide (Ifex) – analog of cyclophosphamide

Therapeutic use – testicular cancer, sarcomas, lymphomas, lung and cervix carcinomas.

Clinical toxicity – as cyclophosphamide, severe urinary tract toxicity.

Melphalan (Alkeran)

Absorption and administration – given orally and intravenously; when given orally, the absorption is incomplete – 20-50% of drug excreted with stool.

Therapeutic use – orally – for treatment of multiple myeloma; intravenously – broad range of neoplastic diseases.

Clinical toxicity – similar to other alkylating agents.

Chlorambucil (Leukeran)

Absorption and administration - given orally and intravenously; good oral absorption and distribution.

Therapeutic use – treatment of chronic lymphocytic leukemia, primary macroglobulinemia.

Clinical toxicity – moderate myelosuppression, amenorrhea, lung fibrosis, seizures, dermatitis, hepatotoxicity, may cause secondary leukemia in long-term use.

Ethylenimines and Methylmelamines

Triethylenemelamine (TEM)

Thiotepa

Altretamine

Absorption and administration – given intravenously; almost fully metabolized (less than 10% appears in urine).

Therapeutic use – Thiotepa – in treatment of bladder cancer, others – ovarian cancer.

Clinical toxicity - similar to other alkylating agents.

Alkyl sulfonates

Busulfan (Myleran)

Absorption and administration – administrated orally, good absorption and distribution; almost fully excreted by urine as methanesulfonic acid.

Therapeutic use – treating of chronic leukemias

Clinical toxicity – myelosuppression, nausea, vomiting, impotence, amenorrhea, fetal malformations, may cause secondary leukemias. In high doses – lung fibrosis, CNS disturbances, seizures, cataract, Addison's disease.

Nitrosoureas

Carmustine

Absorption and administration – unstable in aqueous solutions and body fluids (disappears from plasma in 15-90 min after infusion); administrated intravenously.

Therapeutic use – brain tumors, gastrointestinal cancer, Hodgkin's disease.

Clinical toxicity - myelosuppression, nausea, vomiting, in high doses – lung fibrosis, renal and hepatic failure.

Lomustine, Semustine – analogs of Carmustine

Absorption and administration – given orally, high bioavailability

Therapeutic use – as Carmustine

Clinical toxicity – similar to Carmustine

Triazenes

Dacarbazine

Absorption and administration – administrated intravenously; activated in liver; rapidly disappears from plasma (20 min after infusion); almost ½ of the dose excreted intact with urine.

Therapeutic use – treatment of malignant melanoma, Hodgkin's disease, sarcomas.

Clinical toxicity – major are nausea and vomiting, also seen myelosuppression, usually moderate, flulike syndrome, alopecia, neuro- and hepatotoxicity.

II.   ANTIMETABOLITES

Folic acid analogs

Methotrexate

Mode of action – inhibition of dihydrofolatreductase and folat-depending enzymes, interfering with synthesis of DNA and RNA precursors.

Mechanism of tumor cells resistance to antifolates – 1) impaired transport of drug into the tumor cell; 2) production of altered forms of dihydrofolatreductase with decreased affinity to the inhibitor; 3) increased concentration of dihydrofolatreductase into the cell; 4) decreased activity of folat-depending enzymes.

Absorption and administration – administrated orally and intravenously; good oral absorption in doses less than 25mg\m2, in higher doses absorption is much poorer. Wide distributed in the body, approximately 50% of drug are connected to plasma proteins, and may be displaced by some other drugs, such as sulfonamides, salicylates, tetracyclines, chloramphenocol and phenytoin, which can thus to increase toxicity of methotrexate. About 90% of the drug is excreted with urine during 48 hours after intake.

Therapeutic use – psoriases, rheumatoid arthritis, acute lymphatic leukemia in children, chorioncarcinoma, osteosarcomas, non-Hodgkin's lymphomas, various carcinomas.

Clinical toxicity – myelosuppression, nephrotoxicity (uses with leucovorine), alopecia, dermatitis, interstitial pneumonitis, teratogenesis.

Pyrimidine analogs

Mode of action – inhibits biosynthesis of pyrimidine nucleotides or change them to form that interfere with normal cellular functions, such as synthesis or function of nucleic acids.

Mechanisms of tumor cells resistance – 1) decreased activity of enzymes necessary for activation of the drugs; 2) production of altered enzymes, which are not inhibited by this group of agents.

Fluorouracil (5-FU)

Absorption and administration – administrated intravenously, since oral absorption is incomplete; dose do not have to be reduced in patients with hepatic diseases.

Therapeutic use – carcinomas of breast and GIT, ovary, cervix, bladder, prostate, pancreas, etc.

Higher response is seen then 5-FU is used in combination with other cytotoxic agents (such as cyclophsphamide, methotrexate, cis-platinum), because of synergetic effects.

Clinical toxicity – anorexia, nausea, stomatitis, diarrhea, GIT ulceration, myelosuppression, alopecia; dermatitis, neurological and cardiac toxicity also seen.

Cytarabine (Cytosar-U)

Absorption and administration – administrated orally, intravenously and subcutaneously; good absorption and distribution; only approximately 10% of the drug is excreted with urine, mostly metabolized in the body.

Therapeutic use – acute leukemias in children and adults, non-Hodgkin's lymphomas.

Clinical toxicity – myelosuppression, GIT disturbances, stomatitis, pneumonitis; neurotoxicity seen in high doses.

Purine analogs

Mode of action – inhibits biosynthesis of purine nucleotides or change them to form that interfere with normal cellular functions, such as synthesis or function of nucleic acids.

Mechanism of tumor cells resistance – 1) decreased drug transport; 2) increased rate of degradation of the drug; 3) changes in production or activities of intracellular enzymatic systems.

6-Mercaptopurine (Purinetol)

Absorption and administration – given intravenously, sine oral absorption is incomplete (only 5%) and bioavailability is variable. Often given with allopurinol, which increase activity of the mercaptopurine and decrease possible hyperuricemia during treatment by inhibition of his inactivation by xantine-oxydase.

Therapeutic use – leukemias, lymphomas

Clinical toxicity – myelosuppression, nausea, vomiting, jaundice, rare – GIT disturbances.

Azathioprine – derivative of mercaptopurine. Effects and use are similar.

Thioguanine (6-TG)

Absorption and administration – available orally, though oral administration is incomplete. As mercaptopurine, may be given with allopurinol for the same purposes.

Therapeutic use – treatment of acute leukemia.

Clinical toxicity – myelosuppression, GIT disturbances.

Pentostatine (Nipent)

Absorption and administration – administrated intravenously; almost complete renal elimination (appropriate reduction of dose recommended for patients with renal disorders)

Therapeutic use – treatment of hairy-cell leukemia

Clinical toxicity - myelosuppression, GIT disturbances, impaired liver functions, skin rashes, in high doses neuro- and nephrotoxicity.

III.   NATURAL PRODUCTS

Antimitotic drugs (Alkaloids)

Mode of action – cell-cycle specific agents that block cell mitosis. They are able to bind specifically to tubulin and thus inhibit normal activity of microtubuli in the mitotic apparatus of the cell.

Mechanisms of tumor cell resistance – 1) mutations in the tubulin, that prevent binding of the drug; 2) increased levels of P-130 glycoprotein, a membrane efflux pump that transport the drug out from cell (Ca-channels blockers, such as verapamil, inhibit this type of resistance).

Vincristine (Oncovin), Vinblastine (Velban), Vinorelbine (Navelbine) – alkaloids of Vinca rosea.

Absorption and administration - administrated intravenously, Vinorelbine also orally (oral bioavailability 30%); metabolized by liver, excreted with bile.

Therapeutic use – vinblastine – testicular cancer, Hodgkin's disease, lymphomas, carcinoma of breast, chorioncarcinoma; vincristine – leukemias, Hodgkin's disease, lymphomas, Wilms tumor, neuroblastoma, brain tumors, carcinomas of breast, bladder, reproductive system.

Clinical toxicity – myelosuppression, neurotoxicity, alopecia, GIT disorders, nausea, vomiting, rare – cardiac toxicity; local effects – phlebitis.

Paclitaxel (Taxol)

Absorption and administration – administrated intravenously; metabolized by P-450 in liver (patients with hepatic disorders need the dose reduction), only 10 % excreted with urine.

Therapeutic use – mostly - ovarian and breast cancer, also carcinomas of neck, head, bladder.

Clinical toxicity – myelosuppression, brady- or tachycardia, in high doses – neurotoxicity.

Epipodophyllotoxins

Mode of action – cell-cycle specific agents that block cell mitosis; binding to topoisomerase and DNA, that result in DNA breaks (cells most sensitive in phases S and G-2)

Mechanisms of tumor cell resistance – 1) mutations of topoisomerase; 2) increased levels of P-130 glycoprotein.

Etoposide (Vepeside)

Absorption and administration – administrated orally and intravenously, in oral intake absorption in about 50%, approximately 40% is excreted with urine (patients with renal disorders need dose reduction).

Therapeutic use – testicular cancer, carcinomas of lung and breast, lymphomas, leukemias.

Clinical toxicity – myelosuppression, alopecia, nausea, vomiting, stomatitis, diarrhea, allergy, in high doses – hepatotoxicity.

Teniposide (Vumon, VM-26) – analog of Etoposide

Absorption and administration – administrated intravenously, 45% excreted with urine. Anticonvulsants (dilantine) increase hepatic metabolism of teniposide. Dosage have to be reduced in patients with impaired renal function.

Therapeutic use – treatment of acute lymphoblastic leukemia

Clinical toxicity - myelosuppression, alopecia, nausea, vomiting.

Antitumor Antibiotics

Dactinomycine (Actinomycine D)

Mode of action – bind to DNA and cause blockage of transcription of DNA; in addition, the drug cause break of DNA by free radicals production.

Absorption and administration – give intravenously; metabolism is minimal, excreted with urine and bile, does not cross Blood Brain Barrier (BBB).

Therapeutic use – treatment of rhabdomyosarcoma, Wilms tumor in children, lymphomas, for immunological response inhibition, particularly in renal transplantation.

Clinical toxicity – nausea, vomiting, myelosuppression, diarrhea, glossitis, stomatitis, alopecia, erythema, local effects – in extravasation – severe inflammation in the site of infusion.

Daunorubicin, Doxorubicin, Idarubicin

Mode of action – affecting DNA and RNA synthesis, cause break of DNA strains by producing free radicals.

Absorption and administration – administrated intravenously; rapid distribution, do not cross BBB; metabolized in liver (need to reduce the dose in patients with impaired hepatic function).

Therapeutic use – Daunorubicin – treatment of acute leukemias; Doxorubicin – leukemias and lymphomas, breast cancer; Idarubicin – the same.

Clinical toxicity – myelosuppression, alopecia, GIT disorders, cardiac toxicity.

Bleomycin

Mode of action – causing fragmentation of DNA by producing of free radicals.

Absorption and administration – administrated intravenously; 2/3 of the given dose are excreted with urine (doses should be reduced in presence of impaired renal function).

Therapeutic use – testicular and ovarian cancers, lymphomas, carcinomas of neck and head.

Clinical toxicity – skin toxicity – hyperkeratosis, hyperpigmentation, erythema, ulceration; pulmonary toxicity – cough, rails, infiltrates, lung fibrosis; nausea, vomiting, headaches, hyperthermia.

Mitomycine

Mode of action – alkylation of DNA; in addition – cause breakage of DNA and chromosomes.

Absorption and administration - administrated intravenously; widely distributed in the body, inactivated by metabolism, only 10% is excreted with urine and bile, do not cross BBB.

Therapeutic use – carcinomas of cervix, colon, rectum, breast, lung head and neck.

Clinical toxicity – myelosuppression, nausea, vomiting, in high doses – neurotoxicity, pulmonary infiltration, hemolysis, nephrotoxicity.

Enzymes

L-Asparaginase

Mode of action – catalyses the big amounts of intracellular asparagine, causing deficit in this substance and cell death.

Mechanism of tumor cell resistance – increased capacity of tumor cell to produce asparagine.

Absorption and administration – given intravenously; volume of distribution is approximately the plasma volume.

Therapeutic use – treatment of acute lymphatic leukemia

Clinical toxicity – hypersensitivity and anaphylaxis, abnormalities of clotting factors, immunosuppression, pancreatitis, hyperammonemia.

IV.   MISCELLANEOUS CYTOTOXIC AGENTS

Platinum coordination complexes

Cisplatin, Carboplatin

Mode of action – can react with DNA, forming cross-links, this inhibit DNA replication and transcription and cause DNA breaks.

Absorption and administration - given intravenously; more than 90% is covalently bound to plasma proteins, excreted mostly with urine.

Therapeutic use – ovarian and testicular cancer, carcinomas of head and neck, bladder, lung, etc.

Clinical toxicity – lower in Carboplatin; nephrotoxicity, alopecia, tinnitus, hearing loss, electrolyte disturbances, myelosuppression, anaphylaxis.

Hydroxyurea

Mode of action – inhibits biosynthesis of DNA; specific for S-phase of cells.

Absorption and administration – given orally; rapid oral absorption and distribution; cross BBB; 80% excreted with urine.

Therapeutic use – treatments of myeloproliferative diseases (lymphomas), melanoma, carcinomas of head and neck, cervix.

Clinical toxicity – myelosuppression, GIT disorders, dermatitis.

Procarbazine

Mode of action – methylating DNA; cause chromosomal damage; also products free radicals.

Absorption and administration – must undergo activation by P-450; given parenteral and orally, absorbed almost completely from GIT, metabolized in liver, induction of P-450 by other drugs increase metabolism of procarbazine; 70% excreted with urine.

Therapeutic use – treatment of Hodgkin's disease, lymphomas, brain tumors.

Clinical toxicity – myelosuppression, nausea, vomiting, GIT disorders, CNS depression (use of sedative drugs should be avoided), hypertension, immunosuppression, may cause acute leukemia in long-term use.

Mitotane

Mode of action – not clear; affecting adrenocortical cells (normal and neoplastic).

Absorption and administration – given orally; 40% is absorbed from GIT; wide distribution, 60% excreted with stool.

Therapeutic use – adrenocortical carcinoma

Clinical toxicity – anorexia, nausea, dermatitis, adrenal insufficiency (corticosteroids should be given during the treatment).

V.   BIOLOGICAL RESPONSE MODIFIERS

Interleukin-2 (IL-2)

Mode of action – induce T-cells response cytolytic to tumor cells

Absorption and administration – administrated by continuous infusion because of short half-life (13 min).

Therapeutic use – leukemias, melanoma, renal cancer.

Clinical toxicity – hypotension, arrhythmia, peripheral edema, elevated liver functions, nausea, vomiting, diarrhea, confusion, fever, anemia, thrombocytopenia.

Granulocyte Colony-Stimulating Factor (G-CSF, Filgrastim)

Mode of action – enhance the mobilization of stem cells from bone marrow to peripheral blood.

Absorption and administration – oral and parenteral administration.

Therapeutic use – prophylaxis of chemotherapy-induced neutropenia.

Clinical toxicity – bone pain (resulting from an expansion of cells and increased blood flow in the medullar space).

Granulocyte/Macrophage Colony-Stimulating Factor (GM-CSF, Sargramostim)

Mode of action – enhances antibody-dependent cellular cytotoxicity by stimulation of macrophages, neutrophils and eosinophils. More potent than G-CSF.

Absorption and administration – parenteral administration.

Therapeutic use - prophylaxis of chemotherapy-induced myelosuppression, bone marrow recovery after transplantation.

Clinical toxicity – bone pain, fever, myalgia; rare – tachycardia, hypotension, flushing, pericarditis.

DRUGS ACTING ON THE PARASYMPATHETIC NERVOUS SYSTEM

Background

The neurotransmitter of parasympathetic nerves is acetylcholine. It has two types of receptors – nicotinic and muscarinic. The nicotinic receptors are found at the neuromuscular junction, autonomic ganglia and in the CNS. The muscarinic receptors are found mostly on autonomic effector cells innervated by postganglionic parasympathetic nerves.

A.   MUSCARINIC RECEPTOR AGONISTS

Two types – acetylcholine and other choline esters - methacholine, carbachol, bethanechol

Endogenous and synthetic cholinomimetic alkaloids – pilocarpine, muscarine, arecoline (found in betel nuts)

Choline esters

1.   Acetylcholine has diffuse effects and is rapidly broken down and so is not often used`` in clinical situations.   The synthetic compounds are longer acting and more selective.

2.   Effects

a.   Cardiovascular system – vasodilatation and hypotension, negative chronotropic effects (decreased pulse), negative inotropic effects (decrease in force of contraction) and decrease in the rate of conduction in the sinoatrial and atrioventricular nodes; the hypotension is usually followed by a reflex tachycardia. Only acetylcholine and methacholine have significant effects on the heart.

b.   GIT – increased muscle tone and peristalsis, increased secretions – expressed as nausea, belching, vomiting, abdominal cramps and bowel movements

c.   Urinary tract – decrease bladder capacity and increase peristalsis

d.   Glands – salivary, lacrimal, sweat, tracheobronchial – increased secretions

e.   Other – bronchoconstriction, pupil constriction

3.   Routes of administration – oral, subcutaneous, topical to the eye (intravenous or intramuscular administration is less selective with more toxic effects)

4.   Uses

a.   GIT – bethanechol is used for postoperative gastric atony and gastroparesis, and in some cases of Hirschprung's disease and gastroesophageal reflux.`

b.   Bladder – bethanechol is used in neurogenic bladder and in urinary retention after surgery or childbirth

c.   Eyes – for miosis and in glaucoma

d.   Methacholine is used to diagnose asthma by bronchial challenge

5.   Side effects

a.   Flushing, sweating, headache, salivation, problems with visual accommodation, belching and abdominal cramps

b.   Large amounts can cause complete heart block.

6. Contraindications – hyperthyroidism, asthma, peptic ulcer disease, coronary artery disease

Cholinomimetics

1.   Effects

a.   Cardiovascular system – hypotension and bradycardia

b.   Eye – miosis with initial increase in intraocular pressure and then a sustained (up to 24 hours) decrease, spasm of accommodation (which lasts about 2 hours)

c.   Exocrine glands – excess sweating, salivation, respiratory tract secretions, etc.

d.   Increased tone of smooth muscles of GIT and urinary tract

2.   Routes of administration – oral and topical to the eye

3.   Uses of pilocarpine

a.   Dryness of mucus membranes – as in Sjogren's syndrome or after head/neck irradiation

b.   Glaucoma

4.   Side effects - nausea, hiccups, vomiting, weakness

5.   Toxic effects

a.   Excess salivation, sweating and tearing, nausea, vomiting, diarrhea, abdominal pain, bronchospasm, headache, bradycardia, hypotension and shock

b.   Treatment is atropine.

B.   INHIBITION OF MUSCARINIC RECEPTORS

Mode of action

Prevent acetylcholine from binding to the muscarinic receptor by competing with it for receptors; much less effect on nicotinic receptors

General uses of muscarinic antagonists

1.   GIT – irritable bowel syndrome to decrease motility, reduction of excess salivation

2.   Ophthalmology for mydriasis (pupil dilatation) and cycloplegia (inhibition of accommodation)

3.   Respiratory tract – reduce secretions and cause bronchodilatation

4.   Cardiovascular – limited uses in conditions where there is increased vagal tone (myocardial infarction, AV blocks)

5.   CNS – to treat Parkinsonism and the extrapyramidal effects of neuroleptic drugs; as antiemetic in cases of motion sickness

6.   Used in general anesthesia to decrease respiratory secretions and vagal effects

7.   To treat organophosphate poisoning (chemical warfare or pesticides)

Examples

Belladonna alkaloids

1.   Examples are atropine (most well-known) and scopolamine

2.   Atropine effects receptors in a dose-dependent fashion.

a.   First hit are sweating, salivation and respiratory secretions.

b.   With increased dose, pupil dilates, heart rate increases and accommodation is inhibited.

c.   Last affected are the bladder and GIT.

3.   Effects

a.   CNS - atropine has no real effect on the CNS; scopolamine passes the blood-brain barrier and has CNS effects such as sleepiness, loss of memory, fatigue, decreased REM sleep and euphoria.

b.   Cardiovascular

(1)   is transient bradycardia, but the main effect is tachycardia

(2) Atropine counteracts vagal effects caused by carotid stimulation, pressure on the eyeballs, peritoneal stimulation, etc.

(3) Used to treat second degree AV block

c. Respiratory tract – inhibit respiratory secretions and bronchoconstriction, reduce laryngospasm

d. GIT – decrease gastric motility and secretions (especially saliva)

e. Other effects include decreased sweating, decreased tone of urinary system

f. Although atropine crosses the placenta, it does not seem to have any adverse effects on the fetus.

4. Administration can be oral, parenteral or topical to the eye or mucus membranes.

5. Uses

a. Antidote to excess muscarinic cholinergic stimulation

b. To relieve extrapyramidal symptoms of neuroleptic treatment and Parkinsonism

c. Scopolamine in a transdermal patch is useful in treating motion sickness.

6.   Toxic effects - tachycardia, dry mouth, mydriasis; at higher doses, see headache, palpitations, blurred vision, ataxia, hallucinations, psychoses, convulsions, delirium and coma

7.   Absolute contraindication is atropine with closed angle glaucoma. ("atropine with glaucoma – no diploma")

Quaternary ammonium compounds

1.   Most well-known is ipatroprium bromide; other examples are methantheline, methscopolamine and propantheline

2.   Not effective orally, in the eye or in the CNS; in general longer acting than the belladonna alkaloids

3.   Ipatroprium is administered in inhalation solutions and systemic effects are minimal.

4.   Also cause inhibition of ganglionic (nicotinic) receptors and so have nicotinic side effects as well (such as orthostatic hypotension and impotence).

5.   Ipatroprium is used in the chronic (not acute) treatment of asthma and COPD; unlike other anticholinergics it does not decrease the activity of the cilia of the respiratory tract

6.   Toxic effects – respiratory muscle paralysis

Tertiary amine compounds

Uses

a.   Ophthalmological use; less long acting than atropine or scopolamine - examples are homatropine hydrobromide, cyclopentolate hydrochloride and tropicamide.

b.   To counteract extrapyramidal effects and Parkinsonism , cross the blood-brain barrier – benztropine mesylate and trihexyphenidyl hydrochloride

c.   As antispasmodics in the urinary bladder – oxyphencyclimine hydrochloride, flavoxate hydrochloride

C.   ANTICHOLINESTERASE AGENTS

1.   Cause prolonged action of acetylcholine (after its release from nerve endings) by inhibiting the enzyme that breaks it down (acetylcholinesterase)

2.   Examples used in therapeutic medicine are physostigmine, neostigmine, pyridostigmine.

3.   Effects – as in muscarinic agonists

4. Can be given orally, subcutaneously, topically and parenterally

5. Uses

a. Glaucoma

b. To increase bladder and GIT motility – best is neostigmine

c. Myasthenia gravis – for treatment (usually pyridostigmine or neostigmine) and diagnosis (edrophonium)

d. Alzheimer's disease – tacrine helps with symptoms, but does not stop or slow down the process of deterioration

e. As pesticides (parathion, malathion) and chemical warfare (sarin, soman, tabun)

f. Pyridostigmine can be used as prophylaxis against chemical warfare.

4. Toxic effects – iatrogenic, inadvertent or intentional

a. Muscarinic effects - pinpoint pupils, congestion of conjunctiva, pain of eyes and forehead, runny nose, bronchoconstriction, diarrhea, nausea, vomiting, abdominal pain, excess salivation and sweating, bradycardia and hypotension

b. Nicotinic effects – muscle weakness and fasciculations, paralysis

c.   CNS effects – confusion, ataxia, areflexia, convulsions, coma

d.   Treatment – atropine plus pralidoxime (acts at the neuromuscular junction; atropine does not), along with supportive measures

e.   Tacrine causes reversible elevations of liver enzymes.

D.   NICOTINIC RECEPTOR AGONISTS – AUTONOMIC GANGLIA

Nicotine

1.   Stimulates the receptors at first; the persistent depolarization causes desensitization and inhibition afterward, which is the longer lasting effect.

2.   Acts on both sympathetic and parasympathetic receptors – thus its effects can be variable.

3.   Causes dependence in users (i.e. in cigarette smokers), and so smoking is very difficult to stop.

4.   Effects

a.   CNS – stimulation causes tremors and even convulsions; the subsequent depression can be fatal.   Also causes vomiting, centrally via the emesis zone of the brain and peripherally from vagal stimulation

b. Cardiovascular – tachycardia, hypertension and vasoconstriction (sympathetic)

c. GIT – increased motility and secretions (parasympathetic)

5. Routes of administration – oral, transdermal, in cigarettes and pipes

6. Uses – nicotine transdermal patch or gums are used to help smokers quit smoking. They are less potent than cigarettes and doses can gradually be decreased.

7. Toxic effects – nausea, vomiting, abdominal pain, excess salivation, diarrhea, mental confusion, disturbances of vision and hearing, headache, dizziness and respiratory failure. Treatment is to induce vomiting and give activated charcoal, together with supportive measures.

8. Nicotine withdrawal symptoms – irritability, restlessness, anxiety, depression, impaired concentration, bradycardia, increased appetite

E. NICOTINE RECEPTOR ANTAGONISTS – AUTONOMIC GANGLIA

1. Trimethaphan competes with acetylcholine for the receptor; hexamethonium blocks the open channel.

2. Effects may be on sympathetic or parasympathetic manifestations, according to which receptor is predominant.

a. Sympathetic blockade of blood vessels causes vasodilatation and hypotension.

b. Cholinergic blockade at the heart, eyes, GIT, salivary glands, sweat glands and bladder causes tachycardia, mydriasis and cycloplegia, reduced GIT motility, dry mouth, lack of sweating and urinary retention.

3. Uses

a. Pentolinium is used in general anesthesia to keep blood pressure down.

b. Trimethaphan is used to control hypertension in patients with acute dissection of aortic aneurysm before surgery and to control hypertension during other major surgery.

4. Toxic effects – cycloplegia, urinary retention, impotence, constipation, syncope, paralytic ileus and orthostatic hypotension

Mechanism of action/Effect:

Ganglionic blocking agent ; prevents stimulation of postsynaptic receptors by competing with acetylcholine for these receptor sites; additional effects may include direct peripheral vasodilation and release of histamine. Trimethaphan's hypotensive effect is due to reduction in sympathetic tone and vasodilation, and is primarily postural. Cardiac output may increase in patients with cardiac failure or decrease in patients with normal cardiac function.

SOURCE:

http://www.drugs.com/MMX/Trimethaphan_Camsylate.html

F.   NICOTINE RECEPTOR ANTAGONISTS – NEUROMUSCULAR JUNCTION

Two types – depolarizing agents (succinylcholine) and stabilizing (competitive) agents (curare, atracurium, pancuronium, tubocurarine)

1.   Effects

a.   Paralysis of skeletal muscles; succinylcholine also causes fasciculations

b.   No CNS effects – can't pass blood-brain barrier

c.   Various effects on autonomic ganglia – tubocurarine causes hypotension and tachycardia, other stabilizing agents less so; succinylcholine usually has no significant effects.

d.   Tubocurarine releases histamine with the effects of wheals, bronchospasm, hypotension and increased saliva and bronchial secretions.

2. Oral absorption is very poor. Agents are used intramuscularly or intravenously.

3. Uses

a. Main use is muscle relaxation in surgery, enabling lower doses of inhalational anesthetics.

b. Also used to achieve muscle relaxation in electroconvulsive therapy in psychiatry

4. Toxic effects

a. Succinylcholine causes hyperkalemia – especially in patients with electrolyte imbalances, congestive heart failure or who are taking digitalis or diuretics

b. Persistent apnea – i.e. failure to wake up promptly after anesthesia

c. Malignant hyperthermia – hyperthermia, metabolic acidosis, muscle contractures and tachycardia

5. Contraindications to use of succinylcholine – trauma, burns, non-traumatic rhabdomyolysis, paraplegia or quadriplegia, muscular dystrophy

6. Interactions

a.   General anesthetics – synergism with competitive agents

b.   Antibiotics – aminoglycosides, clindamycin, polymyxin B and tetracycline also produce neuromuscular blockades

c.   Calcium channel blockers – enhance neuromuscular blockade by both types of agent

d.   Anticholinesterases – reverse the effects of tubocurarine

DRUGS USED IN DERMATOLOGY

General principles

1.   Thirty grams of topical medication will cover the entire adult human body.

2.   Remember, children have more surface area per volume than do adults and so more chance for systemic effects.

3.   Skin permeability is highest on the face, in the intertriginous areas and on the perineum.

4.   Need to take into account that some dermatological disorders alter the properties of the skin and so its response to topical preparations.

5.   Moist skin takes up drug better.

6.   Vehicles for drugs have varying ability to hydrate – the least hydrating are soaks, then lotions (powders in water suspensions), solutions (drugs dissolved in solvents) and creams (oil-in-water

emulsions), with ointments (water-in-oil emulsions) being the least drying. In general, drying preparations are for use in acute inflammation, hydrating preparations for chronic inflammation.

Glucocorticoids

1.　　　　　　　Glucocorticoids have anti-inflammatory and immunosuppressive properties and are widely used in skin disease.

2.　　They may be for systemic or topical use or injected into the skin lesion. They may be given orally, intramuscularly or intravenously for systemic use.

3.　　The glucocorticoids available for topical use are divided into seven classes according to their potency, with Class 1 being the most potent and Class 7 the weakest.

4.　　Effects

5.　　Uses – inflammatory skin diseases

a.　　Topical steroids need only be applied twice daily – more often does not increase efficacy and increases risk of systemic effects.

b.　　Triamcinolone preparations are used for intralesional injection.

c.　　Oral steroids are used in severe skin disease. Long-term therapy is necessary in the collagen vascular diseases (SLE, dermatomyositis, inflammatory vasculitis), sarcoidosis, capillary hemangiomas, bullous diseases (bullous pemphigoid, pemphigus vulgaris, gestational herpes) and pyoderma gangrenosum. Other conditions – acute contact dermatitis, atopic dermatitis, lichen planus, erythema nodosum and exfoliative dermatitis – can be treated with short-term use of steroids.

d.   Best to give oral steroids every other day to reduce side effects. Need to treat with as low a dose as is effective and for as short a time period as possible.

e.   Intramuscular injection is not recommended due to poorer absorption and increased toxic effects.

6.   Side effects

a.   Topical preparations may cause skin atrophy, striae, purpura, teleangiectasias, acne-like skin rash, dermatitis and hypopigmentation. Also common is secondary bacterial and fungal growth. Prolonged use near the eye may cause cataract or glaucoma. The frequent and/or extensive use of high potency preparations can suppress the hypothalamus-pituitary-adrenal axis, especially in children.

b.   Intralesional injection can also cause skin atrophy and hypopigmentation.

c.   Side effects from oral use are many and are dose-dependent. Among the most important are myopathy, cataracts, avascular necrosis, hypertension and psychiatric disorders.

d.   Intravenous use can cause changes in blood pressure, hyperglycemia, changes in potassium balance, psychosis, seizures, anaphylactic shock and even death.

e.   Withdrawal from use of systemic steroids should be gradual, with tapering of the dose every other day or less often. Otherwise, a withdrawal syndrome characterized by arthralgia, myalgia and joint effusions may develop and, in some cases, the original disease flares up.   Acute withdrawal after long-term use may cause adrenal insufficiency, as the endogenous production of steroids has been suppressed and takes time to become active again.

Retinoids

1.   Derivatives of vitamin A that act on the differentiation and proliferation of epithelial cells, sebaceous secretions, inflammation and the immune response

2.   There are three generations – the first is the endogenous retinol, tretinoin and isotretinoin.   The second generation are synthetic analogs – etretinate and aciretin.   The third generation is not yet in clinical use.

3.   Uses

a.   Etretinate is used in psoriasis, especially the inflammatory types, and also in psoriatic arthropathy.   It is usually used together with ultraviolet A or B radiation.

b.   Isotretinoin is useful in acne because it normalizes the keratinization, reduces sebum production and reduces the bacteria involved in acne – Propionibacterium acnes.   It is given orally for moderate to severe acne and is also used in Gram-negative folliculitis, acne rosacea and hidraenitis suppurativa.

c.   Tretinoin reduces hyperkeratinization and so prevents the lesions of acne.   It is used topically.

d.   Isotretinoin and etretinate are used in ichthyoses, leukoplakia and skin cancer – basal cell carcinoma, squamous cell carcinoma, keratoacanthoma and cutaneous T-cell lymphoma. Actinic keratosis is treated by tretinoin or etretinate.

e.   Other conditions that respond to retinoids are sarcoidosis, discoid lupus, Reiter's syndrome, warts, acanthosis nigricans and lichen planus.

f.   Retinoids may also be used in malignant disease of the heads, skin, neck and lung and in premalignant conditions of the skin, mouth and uterine cervix.

4.   Side effects

a.   Tretinoin – local effects on the skin such as peeling, burning, redness and stinging pain; photosensitivity can be avoided by applying the preparation before bedtime.

b.   Isotretinoin mostly affects the skin and the eyes – cheilitis, dry eyes, blepharoconjunctivitis. Staph. aureus may colonize but rarely causes clinical infection. Loss of hair and photosensitivity may be seen. Systemic side effects include hyperlipidemia, muscle and joint pains and headaches. Long-term use may cause hyperostoses and extraskeletal ossification; in children there may be premature closure of the epiphyses.

c.   Etretinate – less effect on the eye; alopecia, disturbed liver functions, sticky, easily-bruised skin and exfoliation are common.

Retinoids are contraindicated in pregnancy – even when used topically – because of the risk of fetal malformations. FDA recommendations allow the use of tretinoin only if potential benefits are greater than the risks. The others are absolutely contraindicated. The critical period is the first three weeks, when the woman usually does not even know she is pregnant, so all women of child-bearing age who use retinoids must be careful in using contraception and should take regular pregnancy tests. Many doctors advise using two methods of birth control, beginning a month before and ending a month after therapy with retinoids (for three years after in the case of etretinate due to its long half-life). Malformations include head-face, thymus, heart and CNS anomalies. Miscarriage is also more common.

Antibiotics used in skin disease

For acne

1.   Acne is assoicated with the anaerobe Propionibacterium acnes.

2.   Topical use of benzoyl peroxide, clindamycin and erythromycin is effective in acne.

3.   Acne rosacea responds to topical metronidazole.

4.   Systemic therapy may be necessary – tetracycline is the most common, but erythromycin, minocycline, clindamycin and trimethoprim-syulfamethoxazole are also used.

For infections of the skin

1.   usual microorganisms involved are S. aureus and S. pyogenes.

2.   Topical treatment is by mupirocin, neomycin, polymyxin B or bacitracin – alone or in combined ointments. These may also be used for prophylaxis.

3.   Systemic treatment is used for impetigo or cellulitis. Erythromycin or penicillinase-resistant penicillin are drugs of choice.

Cytotoxic drugs used in skin disease

1.   Methotrexate – used in psoriasis, pityriasis rubra pilaris, vasculitides; watch for hepatotoxicity

2.   Azathioprine – used instead of steroids in pemphigus and pemphigoid

3.   Fluorouracil (5-FU) – used in actinic keratoses and basal cell carcinoma; can be injected into the lesions of keratoacanthomas, warts and porokeratoses

4.   Cyclophosphamide – used in cutaneous T-cell lymphoma, pemphigus, Behcet's disease, scleromyxedema and cytophagic histiocytic panniculitis; used only in refractory cases due to increased risk of secondary malignancies, especially bladder and myeloproliferative and lymphoproliferative diseases

5.   Cyclosporine – used in severe psoriasis, lichen planus, epidermolysis bullosa acquisita, alopecia, pemphigus, bullous pemphigoid; watch for nephrotoxicity

Other drugs used in skin disease

1.   Dapsone – used in dermatitis herpetiformis, leprosy, pemphigoid diseases, pemphigus, vasculitides; watch for G6PD deficiency

2.   Sulfasalazine – used in psoriasis and pyoderma gangrenosum

3.   Antimalarials (chloroquine, hydroxychloroquine, quinacrine) – used in discoid and systemic lupus erythematosus; watch for retinopathy

4.   Antihistamines – H1blockerscan be given topically for pruritus; H2 blockers for pruritus and warts

Other drugs used in psoriasis

Calcipotriene

1.   Vitamin D analog

2.   Used as topical ointment – not for use on the face or in intertriginous areas

3.   Toxic effect – hypercalcemia

Anthralin

1.   mechanism of action unknown

2.   given topically as paste, often together with salicylic acid

3.   best not to use on face and intertriginous areas

4.   side effects – stains skin and anything else it touches, local irritation

Phototherapy

1.   Ultraviolet light can both damage and heal skin.

2.   Psoralen Ultra-Violet A (PUVA)

a.   combination of psoralen (given as capsules or topically) and ultraviolet A rays

b.   used mostly in vitiligo and psoriasis, but also in cutaneous T-cell lymphomas, atopic dermatitis, alopecia and urticaria pigmentosa

c.   side effects – nausea, blisters, painful red skin; chronic therapy may lead to development of actinic keratoses and squamous cell carcinoma

3.   UVB and coal tar are used in psoriasis. The only serious side effect is folliculitis

4.   Watch for photosensitizing drugs such as phenothiazines, thiazides, sulfonamides, NSAIDs, tetracyclines, benzodiazepines, etc.

DRUGS USED IN OPHTHALMOLOGY

Antibiotics

1.   Topical formulations for use in the eye include the following antibiotics – chloramphenicol, ciprofloxacin, erythromycin, gentamicin, tetracycline, tobramycin, polymyxin B, bacitracin, sulfacetamide, and sulfisoxazole. They are used in corneal ulcers, conjunctivitis, keratitis, endophthalmitis, dacryocystitis, blepharitis, hordeolum (stye). They may also be injected into the vitreous humor (in cases of endophthalmitis) or, if necessary, even given parenterally.

2.   Antiviral agents are also used in the eye. Most commonly used are acyclovir, ganciclovir, foscarnet, vidarabine, idoxuridine and trifluridine. They are used in viral keratitis, herpes zoster ophthalmicus and viral retinitis. They can be given topically, intravitreally or parenterally.

3.   Antifungal agents in use include amphotericin B, nystatin, ketoconazole, flucytosine and others. They may be applied topically, into the conjunctiva or into the vitreous humor.

4.   Antiprotozoal agents – used in infections usually caused by Acanthamoeba and Toxoplasma gondii. Most common used is topical combinations of polymyxin B, bacitracin zinc and neomicine (Neosporin), also imidazole and clindamycin.

Drugs used in treatment of glaucoma

a Cholinergic agonists

Acetylcholine (Miochol), Carbachol (Miostat), Pilocarpine   – useful for miosis in glaucoma treatment and ocular surgery

Side effects – corneal edema, myopia, decreased vision, retinal detachment, brow ache

b Anticholinesterase agents

Physostigmine (Eserine), Demecarium, Echotiophate, Isoflurophate

Side effects – retinal detachment, miosis, cataract, iris cyst, brow ache, stenosis of nasolacrimal system

c Sympathomimetic agents

Dipivefrin (Propine)

Side effects – photosensitivity, conjuctival hyperemia, hypersensitivity

d Adrenergic antagonists

Betaxolol, Carteolol, Timolol, Metipranolol, Levobunolol

Side effects – conjuctival hyperemia

e Osmotic agents

Glycerin, mannitol, isosorbide – for sort-term treatment of acute rise of intraocular pressure.

Available orally, glycerin also topically (but oral preparations is preferred in acute rise of intraocular pressure)

Other agents for ophthalmic therapy

1 Vitamins –

Vitamin A – for treatment of xerophtalmy, keratomalacia, keratoconjuctivitis – topical and systemic usage.

Vitamins C and E – topically for treatment of cataract as antioxidants.

2   Wetting agents and tear substitutes

In use for management of dry eyes, includes artificial tears and ophthalmic lubricants.

Most common used are cellulose polymers (carboxymethyl cellulose, hydroxyethylcellulose, methylcellulose), polyvinyl alcohol and dextran.

CARDIAC GLYCOSIDES, ANTIARRHYTHICS AND DRUGS, USED IN ISCHEMIC HEART DISEASE

CARDIAC GLYCOSIDES

Cardiac glycosides – compounds containing steroid nucleus, lactone ring, and polysaccharide chains, are found in several plants. Most well known of them are Digitalis purpurea, Digitalis lanata and Strophanthus gratus. Now the most commonly prescribed cardiac glycoside is Digoxin because of its convenient pharmacokinetics, routes of administration and availability of measurement in serum.

Mode of action

All cardiac glycosides are highly selective inhibitors of the active transport of Na and K across cell membranes, by binding to specific site of Na-K-ATPase, the enzymatic equivalent of the cellular Na-"pump". This inhibition causes activation of Na-Ca-exchanger and increase of intracellular Ca levels, which interact with contractile proteins of myocardial cells and increasing the contractility of cardiac muscle.

Effects

1)   Positive inotropic effect – increasing of the heart velocity and contractility.

2) Negative chronotropic effect – decreasing of heart automatisity and increasing maximal diastolic resting membrane potential in sino-atrial and atrio-ventricular nodes, causing in this way decreased heart conduction.

3) Vasoconstriction in rapid IV administration - transient effect, via inhibition of Na-K-ATPase and increasing of Ca entry cause effect on vascular smooth muscles.

Pharmacokinetics and dosing

Digoxin's half-life is 48 hours, this permits a once-a-day dosing; then the drug is started, during first week loading dose is given ("digitalisation"), then the treatment is continued with maintenance doses.

Digoxin excreted mainly with urine (most part is unchanged). Tissue reservoir of digoxin is skeletal muscles, so dosing should be based on estimated lean body mass. Monitoring is required during administration (target serum concentration is about 1.0 nanogramm/ml).

Drug interactions of digoxin

Cholestiramine, kaolin-pectin, neomycine, sulfasalasine – decreasing absorption of digoxin

Propafenon, quinidine, verapamile, amiodarone – decreasing distributing volume and renal excretion

Thyroxine – increasing distribution volume and renal clearance

Erythromycine, omeprazole, tetracycline – increasing absorption of digoxin

Captopril, diltiazem, nifedipine, cyclosporine – variable decreasing clearance and distribution volume

Beta-blockers, Ca-blockers, flecainide – decrease heart conduction (enhance negative chronotropic effect of digoxin)

Kaliuretic diuretics – decrease serum and tissue K levels, increase digoxin-induced Na-K-ATP-ase inhibition.

Digoxin toxicity

Digoxin toxicity is more likely to appear in chronic use, in hypokalemic state, and in concurrent use of certain drugs (certain diuretics and heart conduction inhibitors). There are many toxic effects in few body systems:

Psychiatric – delirium, confusion, dizziness, malaise.

Visual – disturbed color vision

Gastrointestinal – anorexia, nausea, vomiting, abdominal pain.

Respiratory – enhanced ventilation

Cardiac – proarrhythmic effects (heart blocks and cardiac arrest)

Antidote therapy – (in presence of toxic effects) – antidigoxin immunotherapy (Fab), administrated intravenously; also advised potassium chloride infusion (even if serum potassium levels are normal).

ANTIARRHYTHMICS

Antiarrhythmic drug therapy forms the mainstay of management for most important arrhythmias.

Aniarrhythmic drugs divided on classes by their electrophysiological effects.

Class I drugs - sodium channel blockers.

 All reduce the maximal rate of depolarization of the action potential and thereby slow conduction. They are divided into classes Ia, Ib, and Ic based on the kinetics of their receptor effects. Drugs with short onset and offset belong to class Ib, those with prolonged effects to class Ic, and those with intermediate effects to class Ia. Class I agents include the older antiarrhythmic drugs (eg, quinidine). They are very effective in suppressing ventricular ectopic beats (VEBs), but to a varying degree, they depress left ventricular performance, and all have been associated with arrhythmogenesis (proarrhythmia effects).

Quinidine - class Ia drug that prolongs action potential and refractoriness. Quinidine syncope is a potentially dangerous effect caused by torsade de pointes; the syncope is idiosyncratic and not predictable. If an initial test dose of quinidine sulfate is tolerated, subsequent oral maintenance dosage is usually 200 to 400 mg q 4 to 6 h. Target plasma concentrations lie between 2 and 6 µg/mL. Elimination half-life (t 1/2) is 6 to 7 h. About 30% of patients develop adverse reactions. GI problems (diarrhea, colic, flatulence) are the most common, fever, thrombocytopenia, and liver function abnormalities also occur. Quinidine is a broad-spectrum agent, effective for the suppression of VEBs and VT and for the control of narrow QRS tachycardias, including atrial flutter and fibrillation. It is one of few drugs that may convert atrial fibrillation to sinus rhythm.

Procainamide - class Ia drug, has much less effect than quinidine on refractoriness. It can be given cautiously IV as 100 mg over 1 to 2 min repeated q 5 min to a usual maximum total dose of 600 mg (rarely up to 1 gm), while monitoring BP and ECG. Oral procainamide has a short elimination t 1/2 (<4 h), necessitating either frequent dosing or the use of sustained-release preparations. Usual oral dosing is 250 mg to 625 mg (rarely up to 1 gm) q 3 or 4 h. Target plasma concentrations are 4 to 8 µg/mL. Almost all patients on chronic therapy (>12 mo) will develop serologic abnormalities (notably a positive antinuclear factor test), and up to 40% will have symptoms and signs of hypersensitivity (arthralgia, fever, pleural effusions). The main metabolite, N-acetylprocainamide (NAPA), also has antiarrhythmic effects and contributes to procainamide's efficacy and toxicity.

Disopyramide - a class Ia agent, produces little change in refractory period. It has powerful anticholinergic effects that play only a minor role in arrhythmia management but are responsible for urinary retention and glaucoma; less serious effects - dryness of the mouth, problems of accommodation, bowel upset. Disopyramide has negative inotropic effects, particularly when used parenterally, and it should be used cautiously in patients with markedly impaired left ventricular function. Oral dosing is usually 100 or 150 mg q 6 h. Parenteral dosing - an initial IV dose of 1.5 mg/kg, followed by an IV infusion of 0.4 mg/kg/h. Disopyramide has an elimination t 1/2 of 5 to 7 h. Target plasma concentrations lie between 3 and 6 μg/mL.

Lidocaine - class Ib agent with substantial first-pass hepatic metabolism, is used only parenterally. It produces minimal myocardial depression and has little effect on the sinus node, atrium, or A-V node but acts powerfully upon His, Purkinje, and ventricular myocardial tissue. It can suppress the ventricular arrhythmias that complicate MI (VEBs, VT) and can reduce the incidence of primary ventricular fibrillation (VF) when given prophylactically in early acute MI.. The usual regimen is 100 mg IV over 2 min followed by a further 50 mg IV 5 min later if the arrhythmia has not reverted. An infusion of 4 mg/min (2 mg/min in those >65 yr) should then be started. If continued for >12 h, toxic levels may be reached. Concomitant β-blocker therapy increases the risk of toxicity, and the lidocaine dose should be halved. Lidocaine's elimination t 1/2 is 30 to 60 min. Target plasma concentrations are 2 to 5 μg/L. Unwanted effects are neurologic (tremor, convulsions). Drowsiness, delirium, and paresthesias may occur with too rapid administration.

Mexiletine - class Ib drug, is an analog of lidocaine with similar electrophysiologic actions but has little or no first-pass hepatic metabolism. Oral dosing is 200 to 250 mg q 8 h. A slow release preparation may be given as 360 mg q 12 h. IV dosing is complicated by mexiletine's large volume of distribution. An initial IV dose of 2 mg/kg given at a rate of 25 mg/min should be followed by a 250-mg infusion over 1 h, a 250-mg infusion over the next 2 h, and a maintenance infusion thereafter of 0.5 mg/min. Mexiletine's elimination t 1/2 is 6 to 12 h, and target plasma concentrations - 1-2 μg/mL. Mexiletine, like lidocaine, has few cardiovascular unwanted effects, but GI (nausea, vomiting) and CNS (tremor, convulsions) effects may limit its acceptability.

Tocainide -class Ib, is congener of lidocaine, with little or no first-pass hepatic metabolism. Oral dosing is 400 mg q 8 h. Elimination t 1/2 is 11 to 15 h, and target plasma concentrations lie between 4 and 10 μg/mL. IV dosing is 750 mg infused over 30 min. Continued IV dosing is possible (1200 mg over 24 h), but early recourse to oral therapy is advised. Tocainide's kinetics, indications for use, and unwanted effects are similar to those of mexiletine, but significant unwanted effects, including agranulocytosis, are more likely.

Phenytoin - class Ib. It was used extensively for arrhythmia management, particularly suppressing the ventricular arrhythmias of digitalis toxicity. It has a long elimination t ½ (22 h). With the advent of newer agents and the decline of digoxin toxicity (which may better be treated by digoxin immune fab [Digibind®]), it has little continuing antiarrhythmic role.

Class Ic drugs - are among the most powerful antiarrhythmics but have been associated with a significant risk of proarrhythmia and depression of cardiac contractility. At present, the Ic drugs are used in these latter patients only when the arrhythmia has proved unresponsive to other therapy.

Flecainide - class Ic. By a profound effect on the sodium channel, conduction is markedly slowed but refractoriness is little affected. The proarrhythmia risk of the Ic agents is high. Both flecainide and encainide were associated with an increased mortality (arrhythmogenic). Flecainide is given orally 100 mg q 8 or 12 h. Elimination t 1/2 is 12 to 27 h, and target plasma concentrations are 0.2 to 1μg/mL. It usually is well tolerated, but blurred vision and paresthesia are occasionally reported.

Encainide - class Ic, has similar efficacy and toxicity to flecainide. Unlike flecainide, encainide has at least 3 active metabolites, which are variably formed depending on genetically inherited degradation pathways. Oral maintenance is 75 to 150 mg/day in divided doses. Its long biologic t ½ relies on the active metabolites, so that plasma concentrations of the parent compound (elimination t ½ is about 3 h) are of limited clinical value.

Propafenone - class Ic, has effects similar to those of flecainide and similarly proarrhythmic. Despite low and variable bioavailability, saturable first-pass metabolism, and variable protein binding, dosing is simple (450 to 900 mg/day in divided doses). Initial doses should be small (<= 150 mg tid), and increases should not exceed 50% of the previous dose. It has an elimination t 1/2 of 6 to 7 h. Target plasma concentrations lie between 5 and 8 μg/mL.

Class II drugs – β - blocking agents

The antiarrhythmic effects of the β-blocking agents (class II) are efective to VF. In general, β- blockers are well tolerated but may depress left ventricular function, particularly in antiarrhythmic doses.

They are contraindicated in bronchospastic airways disease and should be used cautiously in other types of lung disease. GI disturbances, insomnia, and nightmares may occur.

Class III drugs – potassium channels blockers

Interfere with the potassium channel to alter the plateau phase of the action potential and increase refractoriness. Rarely, they can be pro-arrhythmic.

Amiodarone - a powerful class III antiarrhythmic. Amiodarone has a long elimination t 1/2 (>50 days), with substantial delay in onset of action. Initial oral loading doses of 600 to 1200 mg/day for 7 to 10 days. Oral maintenance doses should be the minimum consistent with arrhythmia control, ideally <= 200 mg/day. Cardiovascular toxicity is rare. Amiodarone is too toxic for long-term use, except for serious arrhythmias. Pulmonary fibrosis may be fatal and may occur in up to 5% of patients treated for >5 yr. Other problems - photosensitive dermatitis; hepatic abnormalities; peripheral neuropathy; corneal microdeposits; hypothyroidism and hyperthyroidism. Torsade de pointes is rarely produced by amiodarone. Unless there is no alternative, amiodarone should not be given to children.

Racemic (DL) sotalol - has both class II and III antiarrhythmic properties. Sotalol is given orally as 80 to 160 mg q 12 h. It depresses left ventricular performance and has been associated with arrhythmogenesis.

Bretylium - also possesses class II and class III actions. It may cause marked hypotension and is indicated only for the management of potentially lethal refractory ventricular tachyarrhythmias (intractable VT, recurrent VF). The initial IV dose is 5 mg/kg, followed by 1 to 2 mg/min as a constant infusion; its ventricular effects may be delayed 10 to 20 min. The initial IM dose is 5 to 10 mg/kg, which may be repeated to a total dosage of 30 mg/kg; the maintenance dosage is 5 mg/kg IM q 6 to 8 h. Bretylium usually is effective within 30 min after IM injection. Target plasma concentrations are 1 to 1.5 μg/mL.

Class IV drugs - calcium entry blockers

Calcium antagonists. Nifedipine, like other dihydropyridines, is almost devoid of conduction electrophysiologic effects, but verapamil and diltiazem influence A-V nodal electrophysiology and may alter that of calcium-dependent ischemic cells.

Verapamil - acts principally on the A-V node, slowing conduction. Used IV, it has a special place in the acute management of tachycardias, involve the A-V node. Reportedly, termination rates approach 100% with doses of 5 to 15 mg IV over 10 min. Oral verapamil 40 to 120 mg tid is widely prescribed for arrhythmia prophylaxis, but the substantial first-pass hepatic metabolism may limit its clinical utility.

Diltiazem - has a similar electrophysiologic profile to verapamil. It has a long t $\frac{1}{2}$, but it has no first-pass hepatic metabolism, making it better suited for chronic arrhythmia prophylaxis.

Miscellaneous

Adenosine - a purine nucleoside, which acts through extracellular adenosine receptors to slow or block A-V nodal conduction. It is rapidly metabolized after administration. The dose is 6 mg initially followed by up to 12 mg by rapid IV injection. It can terminate arrhythmias that involve the A-V node. Adenosine may be safer than verapamil for this purpose through its extremely short duration of action, but unwanted effects (dyspnea, chest discomfort, flushing) occur in 30 to 60% of patients. Adenosine may cause bronchospasm and should not be used in asthmatic patients.

NITRATES

Nitroglycerin is a potent smooth-muscle relaxer and vasodilator. Its major sites of action are in the peripheral vascular tree, especially in the venous or capacitance system and on the coronary blood vessels. Even severely atherosclerotic vessels may dilate in areas without atheroma. It also lowers systolic BP, thus reducing myocardial wall tension, a major determinant of myocardial O2 need. Overall, the drug brings myocardial O2 supply and demand into more favorable balance.

Amyl nitrite, an extremely potent vasodilator, may be effective when severe angina is unresponsive to nitroglycerin and complicated by hypertension. An ampul containing 0.18 or 0.3 mL is crushed and its vapor briefly inhaled; the patient should be lying down and in a well-ventilated room. Because of the drug's potency, only 2 or 3 inhalations are required.

Long-acting nitrates are available in oral and cutaneous preparations.

1. Isosorbide dinitrate, orally, is effective within 1 to 2 h, with persistent action for 4 to 6 h. Initial dosage of 10 to 20 mg qid or q 6 h may be increased, depending on response, to 40 mg qid. Sustained release preparations are also available.

2. Pentaerythritol tetranitrate is an oral preparation, effective for about 6 h. The initial dosage of 10 to 20 mg qid or q 6 h may be increased, depending upon response, to 40 mg qid.

3. Cutaneous patches are available in various sizes, each containing a different amount of nitroglycerin; all are designed to provide prolonged therapeutic effect by slow release of drug.

4. Nitroglycerin ointment: The drug is well absorbed from the skin, especially in a moist environment. Dispensed as a 2% preparation (15 mg/2.5 cm [1 in.]), it is applied over the upper torso or arms at 6- to 8-h intervals.

Nitrate tolerance: When plasma concentrations are held constant, tolerance to nitrates develops within 24 h. Severity varies among individuals. Tolerance appears due to sulfhydryl depletion in smooth muscles with resultant reduced activation of cyclic GMP.

HORMONES, VITAMINS, AND MINERALS

PITUITARY GLAND HORMONES

Anterior Pituitary Hormones

1) Adrenocorticotropic hormone (ACTH) - stimulates the adrenal cortex to secrete cortisol and several weak androgens. The ACTH-cortisol axis is central to the response to stress, and in the absence of ACTH, the adrenal cortex atrophies and secretion of cortisol virtually ceases.

2) Alpha- and beta-melanocyte stimulating hormone (MSH) - cause hyperpigmentation of skin and are only significant in disorders in which ACTH levels are markedly elevated (Addison's disease).

3) thyroid stimulating hormone (TSH) - regulates the structure and function of the thyroid gland and stimulates synthesis and release of thyroid hormones. TSH synthesis and secretion are controlled by the hypothalamic hormone, Thyrotropin Releasing Hormone (TRH), and by circulating thyroid hormone from the periphery.

4) Luteinizing hormone (LH) and Follicle Stimulating Hormone (FSH) - in women stimulate ovarian follicular development and

ovulation. In men, FSH is essential for spermatogenesis, and LH stimulates testosterone biosynthesis.

5) Growth Hormone (GH) (somatotropine) – the major actions is stimulation of somatic growth and regulation of metabolism. Somatostatin is the major inhibitor of the synthesis and secretion of GH.

6) Prolactine (PRL) - the major function is regulation of milk production. PRL release also occurs with stress and sexual activity. PRL is the most frequent hormone produced in excess by pituitary tumors.

Posterior Pituitary Hormones

1)   Antidiuretic hormone (ADH, vasopressin) – the major action is to promote water conservation by the kidney. At high concentrations it also causes vasoconstriction. ADH release is stimulated by pain, stress, exercise, hypoglycemia, cholinergic agonists, beta-adrenergic agonists, angiotensin, prostaglandins, etc. Alcohol, alpha-adrenergic agonists, glucocorticoids, etc, inhibit ADH secretion.

2)   Oxytocin - stimulates contraction of uterine smooth muscle cells, and uterine sensitivity to oxytocin increases throughout pregnancy; also stimulates milk delivery from the breast during lactation.

ACROMEGALY AND GIGANTISM

Syndromes of excessive secretion of GH   (hypersomatotropism) nearly always due to a pituitary adenoma of the somatotrophs.

Drugs of treatment:

Bromocriptine mesylate - to 15 mg/day orally in divided doses

Octreotide - a long-acting somatostatin analog has been shown to suppress GH secretion effectively in patients refractory to bromocriptine. Side effects - gastritis, gallstones, cholangitis, jaundice, malabsorption of vitamin B12.

DIABETES INSIPIDUS (DI)

(Central Diabetes Insipidus; Vasopressin-Sensitive Diabetes Insipidus)

A temporary or chronic disorder of the neurohypophyseal system due to deficiency of vasopressin (antidiuretic hormone, ADH) and characterized by excretion of excessive quantities of very dilute (but otherwise normal) urine and by excessive thirst.

Drugs of treatment:

Hormonal therapy: Because vasopressin is a small peptide, it is ineffective when administered orally. Aqueous vasopressin - s.c. or IM in doses of 5 to 10 u. each 6 h or less.

Synthetic vasopressin (Pitressin ®) - bid to qid as a nasal spray, with the dosage and interval designed for each patient.

DDAVP® (desmopressin acetate, 1-deamino-8- d -arginine vasopressin), a synthetic analog of arginine vasopressin, has prolonged antidiuretic activity lasting for 12 to 24 h, administered intranasally, s.c., or IV. Desmopressin acetate is the preparation of choice for both adults and children. The usual dosage in adults is 0.1 to 0.4 mL (10 to 40 µg), with most requiring 0.2 mL/day in 2 divided doses. For children age 3 mo to 12 yr, the usual dosage range is 0.05 to 0.3 mL/day. Overdosage can lead to fluid retention and decreased plasma osmolality, possibly resulting in convulsions in small children. In such instances, furosemide may be used to induce diuresis. Headache may be a troublesome side effect but generally disappears if the dosage is reduced. Infrequently, desmopressin acetate may cause a slight increase in BP. Desmopressin acetate may also be used IV in acute situations.

Lypressin (lysine-8-vasopressin) - a synthetic agent, given by nasal spray as required at 3- to 8-h intervals.

Vasopressin tannate in oil - IM in a dose of 0.3 to 1 mL (1.5 to 5 u.) controls symptoms up to 96 h.

Nonhormonal therapy: At least 2 types of drugs are useful in reducing polyuria: (1) various diuretics, primarily thiazides, and (2) ADH-releasing drugs such as chlorpropamide, carbamazepine, and clofibrate. The thiazides paradoxically reduce urine volume in DI, primarily as a consequence of reducing extracellular fluid (ECF) volume and increasing proximal tubular resorption.

Chlorpropamide (3 to 5 mg/kg orally once or twice/day) causes some release of ADH, also potentiates the action of ADH on the kidney.

Clofibrate 500 to 1000 mg orally bid or carbamazepine 100 to 400 mg orally bid is recommended for adults only.

Prostaglandin inhibitors such as indomethacin (1.5 to 3.0 mg/kg/day orally in divided doses) effective in reducing urine volume perhaps by decreasing renal blood flow and glomerular filtration rate.

THYROID HORMONES

Iodide, ingested in food and water, is actively concentrated by the thyroid gland, converted to organic iodine by peroxidase, and incorporated into tyrosine in intrafollicular thyroglobulin. The tyrosines are iodinated at either one (monoiodotyrosine, MIT) or two (diiodotyrosine, DIT) sites and then coupled to form the active hormones (diiodotyrosine + diiodotyrosine --> tetraiodothyronine [thyroxine, T4]; diiodotyrosine + monoiodotyrosine --> triiodothyronine [T3]). Thyroglobulin, a glycoprotein containing T3 and T4 within its matrix. Lysosomes containing proteases cleave T3 and T4 from thyroglobulin, resulting in release of free T3 and T4.

Physiologic Effects of Thyroid Hormone

(1) increase protein synthesis in virtually every body tissue (T3 and T4 enter cells, bind to discrete nuclear receptors, and influence the formation of mRNA); (2) increase O2 consumption by increasing the activity of the Na+-K+ ATPase (Na pump), primarily in tissues responsible for basal O2 consumption (ie, liver, kidney, heart, and skeletal muscle). The increased activity of Na+-K+ ATPase is secondary to increased synthesis of this enzyme; therefore, the increased O2 consumption is also probably related to the nuclear binding of thyroid hormone. T3 is at least 3 times more active than T4, although T4 itself is biologically active.

HYPERTHYROIDISM

(Thyrotoxicosis; Toxic Diffuse Goiter; Graves' Disease; Basedow's Disease)

Drugs of treatment

Iodine - inhibits the release of T3 and T4 organification of iodine, a transitory effect lasting from a few days to a week ("escape phenomenon"). Iodine is generally not used for routine treatment of hyperthyroidism. The usual dosage is 2 to 3 drops of a saturated potassium iodide solution orally tid or qid (300 to 600 mg/day), or 0.5 gm sodium iodide in 1 L 0.9% sodium chloride solution given IV slowly q 12 h. Complications include inflammation of the salivary glands, conjunctivitis, and skin rashes.

Propylthiouracil and methimazole - decrease organification and impair the coupling reaction.

Propylthiouracil (but not methimazole) in doses >800 mg/day also inhibits the peripheral conversion of T4 to T3. The usual starting dosage for propylthiouracil is 100 to 150 mg orally q 8 h, and for methimazole 10 to 15 mg orally daily. Maintenance doses can be continued for one year or many years depending on the clinical circumstances. Carbimazole is rapidly converted to methimazole. The usual starting dosage is 10 to 15 mg orally q 8 h; maintenance dosage is 10 to 15 mg/day.

Adverse effects - allergic reactions, nausea, loss of taste, and, in <1% of patients, a reversible agranulocytosis.

B-Adrenergic blocking drugs: Symptoms and signs of hyperthyroidism due to adrenergic stimulation may respond to Propranolol.

Radioactive sodium iodine - the treatment of choice in patients >40 yr of age. Dosage of I is difficult to gauge, and the response of the gland cannot be predicted.

HYPOTHYROIDISM

(Myxedema)

The characteristic reaction to thyroid hormone deficiency in the adult.

Drugs of treatment

A variety of thyroid hormone preparations are available for replacement therapy: synthetic preparations of thyroxine, liothyronine (triiodothyronine), combinations of the 2 synthetic hormones, and desiccated animal thyroid. Synthetic preparations of T4 (L-thyroxine) are preferred; the average maintenance dosage is 100 to 125 μg/day orally. Absorption is fairly constant at about 90 to 95% of the dose.

T3 (liothyronine sodium) should not be used alone for long-term replacement because its rapid turnover requires that it be taken bid or tid.

ADRENAL

The adrenal cortex produces androgens, glucocorticoids (cortisol), and mineralocorticoids (aldosterone).

The medulla produces catecholamines (mostly epynephrin)

The principal hormones produced by the adrenal cortex are cortisol (hydrocortisone), aldosterone, and dehydroepiandrosterone (DHEA).

Adults secrete about 20 mg of cortisol, 2 mg of corticosterone (which has similar activity), and 0.2 mg of aldosterone daily. Although considerable quantities of androgens (primarily DHEA and androstenedione) are normally produced by the adrenal cortex, these exert their chief physiologic activity after conversion to testosterone and dihydrotestosterone.

The physiological effects of steroids:

1) Increased levels of serum glucose

2) Increased lipids deposition

3) Increased catalysis of proteins

4) Increased platelets aggregation

5) Kaliuretic effect

6) Na retention effect

7) Increased Ca releasing from the bones

8) CNS stimulation

9) cyclooxygenase (COX) inhibition

10) Inhibition of white blood cell (WBC) and mast cell activity

ADRENAL CORTICAL HYPOFUNCTION (ADDISON'S DISEASE)

Treatment:

Steroids replacement therapy is useful.

Glucocorticoids are divided on classes by the time of their action:

1.   Short-acting : hydrocortisone, prednisone, prednisolone, fluotorton, methylprednisolone.

2.   Moderate-acting : thriamsynolone, paramethasone, fluprednisone.

3.   Long-acting : betamethasone, dexamethasone.

Adverse effects of steroids:

1) leukopenia, thrombocytosis, increased coagulation

2) hypokalemia, hypernatremia, hypercalcemia, hyperglycemia

3) osteoporosis, in children –impaired bones growth rate

4) immunosuppressive state and secondary infections

5) peptic ulcer, obesity, muscular atrophy

6) hypertension, edemas

7) impaired sexual function

8) CNS hyperstimulation (agitation, irritability, delirium)

ADRENAL CORTICAL HYPERFUNCTION (CUSHING'S SYNDROME)

Clinical abnormalities due to chronic exposure to excesses of cortisol or related corticosteroids.

Drugs of treatment

Adrenal inhibitors:

Metyrapone - 250 mg qid orally, increasing to a maximum of no more than 2 gm/day;

Aminoglutethimide - 250 mg bid orally, increasing to a maximum of no more than 2 gm/day;

Mitotane (o,p-DDD) - 0.5 gm qid orally, increasing to a maximum total dose of 8 to 12 gm/day

Ketoconazole - blocks steroid synthesis in all systems

Adverse reactions – the most problematic is inhibition of synthesis of all groups of steroids is the body.

PHEOCHROMOCYTOMA

A tumor of chromaffin cells that secrete catecholamines, causing hypertension.

Drugs of treatment

Usually treatment includes the combination of alpha- and beta-adrenergic blocking agents (phenoxybenzamine 40 to 160 mg/day and propranolol 30 to 60 mg/day, orally in divided doses).

The infusion of trimethaphan camsylate or sodium nitroprusside useful in case of hypertensive crisis.

Metyrosine - orally, may be used alone or in combination with an alpha-adrenergic blocking agent (phenoxybenzamine); the optimally effective dosage is 1 to 4 gm/day in divided doses, should be given for at least 5 to 7 days before surgery.

Labetalol - an agent with both A- and B-adrenergic blocker properties, has been used at a dosage of 200 mg/day orally in divided doses.

DIABETES MELLITUS (DM)

A syndrome characterized by hyperglycemia resulting from impaired insulin secretion and/or effectiveness

Classification and Pathogenesis

Insulin-dependent diabetes mellitus (IDDM, type I DM) is clinically characterized by hyperglycemia and a propensity to DKA. Its control requires chronic insulin treatment. The IDDM is associated with islet cell autoantibodies that selectively bind to beta-cells and cause their destruction.

Non-insulin-dependent diabetes mellitus (NIDDM, type II DM) is characterized clinically by hyperglycemia that is not associated with a propensity to DKA. It is commonly associated with obesity. The concordance rate for NIDDM in monozygotic twins is >90%, and genetic factors appear to be the major determinants of its development.

NIDDM is a disorder in which hyperglycemia results from both an impaired insulin secretory response to glucose and decreased insulin effectiveness (insulin resistance).

Treatment

The maximum acceptable plasma glucose levels are 80-120 mg/dL.

In treating IDDM, chronic insulin therapy is always required.

Insulin Preparations

Preparations of purified porcine insulin, purified bovine insulin, semisynthetic human insulin, and biosynthetic human insulin (all 99% pure, <10 parts per million [ppm] proinsulin) are now available; they have equivalent biologic activities.

Insulin is routinely provided in preparations containing 100 u./mL (U-100 insulin) and is injected s.c. with disposable insulin syringes calibrated for use with U-100 insulin, which are commercially available with maximal capacities of 100 u. (1 mL), 50 u. (0.5 mL), and 30 u. (0.3 mL).

Insulin Preparation	Onset of Action	Peak Action (h)	Duration of Action (h)
Rapid-acting regular	15-30 min	2-4	6-8
Rapid-acting Semilente® (insulin zinc suspension)	(prompt 1½-2 h	4-9	10-16
Intermediate-acting (NPH and Lente®)	1-3 h	6-12	18-26
Long-acting (Ultralente® and PZI)	4-8 h	14-24	28-36

NPH = neutral protamine Hagedorn; PZI = protamine zinc insulin.

GLUCAGON

Glucagon is a polypeptide hormone secreted by alpha cells, found almost exclusively in the pancreatic islets in humans. Glucagon is used to treat severe hypoglycemic reactions due to insulin;

Glucagon is available for injection in vials containing 1 u. (1 mg) or 10 u. (10 mg) of crystalline glucagon. The usual dose of glucagon in adults is 0.5 to 1 u. given s.c., IM, or IV; in children, it is 0.03 u./kg (maximum dose 1 u.). The major side effects are nausea and vomiting.

Immunologic insulin resistance

Occurs in patients receiving insulin longer than 6 months. In this case the dose of insulin should be increased and prednisone treatment initiated for decreasing of immune response.

Oral Hypoglycemic Agents

Oral hypoglycemic agents are in use in treatment of NIDDM.

Biguanides: metformin, phenformin

Effects: 1- decreasing glucose absorption in GIT; 2- inhibit glucagon secretion; 3- stimulation of tissue glycolysis.

Side effects: the major side effect is hypoglucemia; rarely metabolic acidosis may occur.

Sulfonureas: tolbutamide, chlorpropamide, acetohexamide, tolazamide, glyburide, glipizide.

Effects: 1- stimulating insulin secretion; 2- enhancing insulin action in target tissues.

Side effects: hypoglycemia, allergy, jaundice.

THE PRINCIPAL MICRONUTRIENTS (VITAMINS AND MINERALS)

Vitamin A

Principal Sources: as preformed vitamin - fish liver oils, liver, egg yolk, butter, cream; as provitamin carotenoids: dark green vegetables, yellow fruits.

Functions: Photoreceptor mechanism of retina, integrity of epithelia, lysosome stability, glycoprotein synthesis.

Deficiency and Toxicity: deficiency - night blindness, hyperkeratosis, keratomalacia; toxicity – Headache, peeling of skin, hepato-splenomegaly, bone thickening.

Vitamin D

Principal Sources: ultraviolet irradiation, fish liver oils, milk, butter, egg yolk, liver.

Functions: Calcium and phosphorus absorption, mineralization and collagen maturation of bone; tubular reabsorption of phosphorus.

Deficiency and Toxicity: Primary - Rickets (tetany associated), osteomalacia, Metabolic – anorexia, renal failure.

Vitamin E group

Principal Sources: vegetable oil, wheat germ, egg yolk, margarine, vegetables.

Functions: Intracellular antioxidant, scavenger of free radicals.

Deficiency and Toxicity: neurologic damage, creatinuria, ceroid deposition in muscle, RBC hemolysis; toxicity - interferes with enzymes, increased infection.

Vitamin K - Vitamin K1 (phytonadione), Vitamin K2

Principal Sources: vegetables, pork, liver, vegetable oils, intestinal flora after newborn.

Functions: formation of prothrombin and other coagulation factors, normal blood period coagulation.

Deficiency and Toxicity: hemorrhage; toxicity – kernicterus

Essential fatty acids (linoleic, linolenic, arachidonic acids)

Principal Sources: vegetable seed oils (corn, sunflower)

Functions: precursors of prostaglandins, leukotrienes, various hydroxy fatty acids, membrane structure.

Deficiency and Toxicity: growth cessation, dermatosis.

Thiamine (vitamin B1)

Principal Sources: whole grains, meat, nuts, legumes, potatoes.

Functions: carbohydrate metabolism, central and peripheral nerve cell function, myocardial function.

Deficiency and Toxicity: Beri-beri - infantile and adult (peripheral neuropathy, cardiac failure); Wernicke-Korsakoff syndrome.

Riboflavin (vitamin B2)

Principal Sources: Milk, cheese, liver, meat, eggs.

Functions: Many aspects of energy and protein metabolism, integrity of mucous membranes.

Deficiency and Toxicity: angular stomatitis, corneal vascularization, amblyopia, sebaceous dermatosis.

Niacin (Vitamin B3)

Principal Sources: liver, meat, fish, legumes, whole-grain enriched cereal products.

Functions: Oxidation-reduction reactions, carbohydrate metabolism and CNS function.

Deficiency and Toxicity: Pellagra (dermatosis, glossitis, GI and CNS dysfunctions).

Vitamin B6 group (pyridoxine)

Principal Sources: liver, organ meats, fish, legumes.

Functions: many aspects of nitrogen metabolism - transaminations, porphyrin and heme synthesis, tryptophan conversion to niacin; linoleic acid metabolism.

Deficiency and Toxicity: convulsions in infancy, anemias, neuropathy, seborrhea-like skin lesions.

Folic acid

Principal Sources: fresh green leafy vegetables, fruit, organ meats, liver.

Functions: maturation of RBCs, synthesis of purines and pyrimidines.

Deficiency and Toxicity: pancytopenia; megaloblastosis (especially pregnancy, infancy, malabsorption).

Vitamin B12 (cobalamins)

Principal Sources: Liver, meats, fish, eggs, milk.

Functions: Maturation of RBCs; neural function; DNA synthesis, related to folate coenzymes; methionine and acetate synthesis.

Deficiency and Toxicity: Pernicious anemia, some psychiatric syndromes.

Biotin

Principal Sources: Liver, kidney, egg yolk, cauliflower, nuts, legumes/

Functions: Carboxylation and decarboxylation of oxalocetic acid; amino acid and fatty acid metabolism.

Deficiency and Toxicity: dermatitis, glossitis.

Vitamin C (ascorbic acid)

Principal Sources: Citrus fruits, tomatoes, potatoes, cabbage, green peppers.

Functions: Essential to osteoid tissue, collagen formation, vascular function, wound healing.

Deficiency and Toxicity: Scurvy (hemorrhages, loose teeth, tissue gingivitis).

Sodium

Principal Sources: wide distribution - beef, pork, sardines, cheese, green olives, corn bread, etc.

Functions: Acid-base balance, osmotic pressure, blood pH, muscle contractility, nerve transmission, sodium pumps.

Deficiency and Toxicity: hypo- and hypernatremia - confusion, coma.

Potassium

Principal Sources: wide distribution - bananas, prunes, raisins, milk

Functions: muscle activity, nerve transmission; intracellular acid-base balance.

Deficiency and Toxicity: hypokalemia - paralysis, cardiac disturbances; hyperkalemia - paralysis, cardiac disturbances.

Calcium

Principal Sources: Milk and milk products, meat, fish, eggs, cereal products, beans, fruits, vegetables.

Functions: bone and tooth formation, blood coagulation, neuromuscular irritability, muscle contractility, myocardial conduction.

Deficiency and Toxicity: hypocalcemia - neuromuscular hyperexcitability and tetany; hypercalcemia - GI atony, renal failure, psychosis.

Phosphorus

Principal Sources: Milk, cheese, meat, fish, cereals, nuts, legumes.

Functions: Bone and tooth formation, acid-base balance, component of nucleic acids, energy production.

Deficiency and Toxicity: hypophosphatemia - weakness, blood cell disorders, GI tract and renal dysfunction, irritability; hyperphosphatemia - renal failure.

Magnesium

Principal Sources: green leaves, nuts, cereal grains, seafood.

Functions: bone and tooth formation, nerve conduction, muscle contraction, enzyme activation.

Deficiency and Toxicity: hypomagnesemia - neuromuscular irritability; hypermagnesemia – hypotension, respiratory failure, cardiac disturbances.

Iron

Principal Sources: wide distribution - soybean flour, beef, kidney, liver, beans, clams, peaches.

Functions: Hemoglobin, myoglobin formation, enzymes.

Deficiency and Toxicity: deficiency - anemia, dysphagia, enteropathy, decreased work performance, impaired learning ability; toxicity - hemochromatosis, cirrhosis, diabetes mellitus, skin pigmentation.

Iodine

Principal Sources: Seafoods, iodized salt, dairy products.

Functions: Thyroxine (T4) and triiodothyronine (T3) formation and energy control.

Deficiency and Toxicity: deficiency – cretinism, fetal growth and brain development disorders; toxicity – thyrotoxicosis, myxedema.

Fluorine

Principal Sources: wide distribution--tea, coffee, fluoridation of water.

Functions: bone and tooth formation

Deficiency and Toxicity: deficiency - predisposition to dental caries, osteoporosis; toxicity - fluorosis, pitting of permanent teeth; exostoses of spine.

Zinc

Principal Sources: wide distribution--vegetable sources.

Functions: component of enzymes and insulin; skin integrity, wound healing, growth.

Deficiency and Toxicity: growth retardation, hypogonadism, acrodermatitis enteropathica.

Copper

Principal Sources: wide distribution--organ meat, nuts, dried legumes, whole-grain cereals.

Functions: Enzyme component, hemopoesis, bone formation.

Deficiency and Toxicity: deficiency – anemia; toxicity - hepatolenticular degeneration, biliary cirrhosis.

Cobalt

Principal Sources: green leafy vegetables.

Functions: part of vitamin B12 molecule.

Deficiency and Toxicity: deficiency – anemia; toxicity – cardiomyopathy.

Chromium

Principal Sources: wide distribution.

Functions: part of glucose tolerance factor.

Deficiency and Toxicity: impaired glucose tolerance.

Selenium

Principal Sources: meats and other animal products.

Functions: component of glutathione peroxidase.

Deficiency and Toxicity: deficiency – cardiomyopathy; toxicity - loss of hair and nails, dermatitis, polyneuritis.

HYPERTENSIVE DRUGS AND LIPIDS-LOWERING DRUGS

ARTERIAL HYPERTENSION - elevation of systolic and/or diastolic BP, either primary (essential hypertension) or secondary.

Primary or essential hypertension is of unknown etiology.

Whatever the responsible pathogenic mechanisms, they must lead either to increased total peripheral vascular resistance (TPR) by inducing vasoconstriction or to increased cardiac output (CO), or both. The sympathetic nervous system and the renin-angiotensin-aldosterone system have received the most attention by investigators of the pathophysiology of hypertension, since both can increase CO and TPR.

Sympathetic nervous system: there is hyperresponsiveness in the sympathetic nervous system itself or in the myocardium and vascular smooth muscle in the hypertension. Certain manifestation of increased sympathetic nervous activity, is a predictor of subsequent hypertension - hypertensives have higher than normal circulating plasma catecholamines.

The renin-angiotensin-aldosterone system: The juxtaglomerular apparatus (JGA) is involved in volume and pressure regulations. Renin, a proteolytic enzyme formed in the JGA cells, catalyzes the conversion of angiotensinogen (a plasma protein) to angiotensin I, a decapeptide. This inactive product is cleaved by a converting enzyme, mainly in the lung but also in the kidney and brain, to an octapeptide, angiotensin II, which is a potent vasoconstrictor and also stimulates release of aldosterone.

Another important factor for hypertension and other cardiovascular diseases is atherosclerosis, which appears mainly because of impaired lipid balance in the body and hypercholesterolemia.

The major plasma lipids - cholesterol (or total cholesterol [TC]) and the triglycerides (Tgs), do not circulate freely in solution in plasma, but are bound to proteins and transported as macromolecular complexes called lipoproteins.   The major lipoprotein classes-- chylomicrons, very low-density lipoproteins (VLDL), low-density lipoproteins (LDL), and high-density lipoproteins (HDL).

Hypercholesterolemia can result from overproduction of VLDL, increased conversion of VLDL to LDL, or defective clearance of LDL.

Normal Plasma Levels of Total Cholesterol (TC) and triglycerides

The optimal plasma TC for a middle-aged adult is probably $<= 200$ mg/dL (levels between 200 and 240 mg/dL is a borderline, levels $>$ 240 mg/dL is a high-risk).

Drug groups are in use for treatment of hypertension:

1)   Beta-adrenergic blockers

2)   Diuretics

3) Central adrenergic inhibitors

4) Direct vasodilators

5) Ca-blockers

6) Angiotensin-converting enzyme (ACE) inhibitors

7) Lipids – lowering drugs

Step-1 drug: The recommended antihypertensive drug therapy have to be initiated with a diuretic, a β-blocker, a Ca antagonist (Ca channel blocker), or an angiotensin converting enzyme (ACE) inhibitor.

Beta-adrenergic blockers, diuretics, central adrenergic inhibitors, and direct vasodilators are described in corresponding chapters.

CALCIUM ANTAGONISTS

(Calcium Channel Blocking Agents; Calcium Blockers; Calcium Channel Antagonists; Calcium Entry Blockers)

Calcium antagonists inhibit Ca ion movement into cells. Three drugs in this class, diltiazem, nifedipine, and verapamil, are well established in the management of a variety of cardiovascular disorders.

Pharmacology

Although all Ca antagonists belong to the same pharmacologic class, they are heterogeneous compounds with different chemical structures and varying potencies for blocking Ca slow channel. The primary sites of action include cardiac muscle, the cardiac electrophysiologic system, and systemic and coronary arterial smooth muscle cells. The cardiac effects include a decrease in myocardial contractility, a decrease in the rate of the sinus node pacemaker and suppression of atrioventricular (A-V) nodal conduction, and a decrease in resistance in the coronary and systemic arterial systems through vasodilation.

Diltiazem is produced in both oral and IV forms and is approved for the management of angina pectoris and hypertension. It is well absorbed following oral administration, but the bioavailability is only 40 to 50% with extensive first-pass hepatic metabolism. The half-life is 3 ½ to 7 h, and it is most effective in a tid to qid regimen. The sustained-release formulation allows once-daily administration for the management of hypertension.

Nifedipine is the most potent peripheral arterial dilator compared with diltiazem and verapamil. It does not affect A-V nodal conduction, but it may increase heart rate due to reflex baroreceptor stimulation of the sympathetic nervous system from peripheral vasodilation. It is rapidly and well absorbed from the GI tract following oral administration and undergoes significant first-pass metabolism with a bioavailability of about 50%. Metabolic products are inactive. The half-life is 3 to 4 h, most effective when administered tid to qid. The onset of action is about 20 min, with peak effects at 30 min after oral administration. Nifedipine is available as a liquid within a capsule formulation; biting the capsule permits the liquid to be absorbed sublingually, this provides an onset of action in 3 to 5 min.

Verapamil is approved for the treatment of angina, hypertension, and certain supraventricular arrhythmias. Like other Ca antagonists, it is rapidly and almost completely absorbed after oral administration. The half-life is variable between 3-7 h (increasing to 4½ to 12 h with chronic therapy). The onset of action is 30 min with peak effects at 1 to 2 h. It is typically administered tid. A sustained-release formulation allows once-daily administration for the management of essential hypertension. First-pass metabolism is more extensive than for diltiazem and nifedipine, with an initial bioavailability as low as 20%. With chronic administration, liver metabolism decreases and bioavailability increases.

Newer Calcium Antagonists

Recently, a number of 2nd-generation Ca antagonists have become available. Nicardipine, isradipine, nitrendipine, and felodipine possess vascular selectivity with little direct cardiac effect; nimodipine demonstrates selectivity for the cerebral vasculature.

Adverse Reactions

Diltiazem is usually well tolerated and has the lowest incidence of side effects.

Nifedipine causes peripheral vasodilation.

Verapamil's major effects are on the cardiac conduction system and myocardial contractility, potential side effects are high-grade A-V block and heart failure. Constipation is also a common problem and may be particularly troublesome in the elderly.

Adverse Drug Interactions

β -adrenergic blocking agents: verapamil and diltiazem can exacerbate the suppressive effects of certain beta blockers on heart rate, A-V conduction, and myocardial contractility.

Digitalis: Verapamil has been reported to significantly increase digitalis levels in patients on chronic combination therapy.

Cimetidine: plasma levels of nifedipine, diltiazem, and nicardipine are elevated with cimetidine therapy. Patients taking ranitidine have demonstrated smaller and nonsignificant changes in diltiazem and nifedipine levels.

Cyclosporine: significant increases in cyclosporine levels can occur with Ca antagonists.

ANGIOTENSINE-CONVERTING ENZYME INHIBITORS

ACE-inhibitors cause blood pressure decreasing by blocking the process of conversion of the angiotensine-I to angiotensine-II.

Drug	Trade Names	Dose (mg)	Side Effects (common for all the drugs)
Captopril	Capoten	25-300	Rash, cough, angioneurotic edema
Enalapril	Vasotec	2.5-40	hyperkalemia, reversible acute renal failure
Fosinopril	Monopril	10-60	
Lisinopril	Zestril	5-40	in patients with bilateral renal arterial stenosis,
Ramipril	Altace	2.5-10	nephrotic syndrome, glomerulonephritis.

LIPIDS-LOWERING DRUGS

Drugs lower elevated lipid levels by several mechanisms: (1) Bile acid sequestrants (cholestyramine and colestipol) and 3-hydroxy-3-methylglutaryl coenzyme A (HMG-CoA) reductase inhibitors (lovastatin, pravastatin, simvastatin, and fluvastatin) stimulate the clearance of LDL via receptor-mediated mechanisms. (2) Nicotinic acid (niacin) reduces the rate of synthesis of VLDL, the precursor of LDL. (3) Fibric acid derivatives (gemfibrozil, clofibrate, fenofibrate and bezafibrate) accelerate the clearance of VLDL. (4) Probucol, a bis-phenol, stimulates the clearance of LDL via nonreceptor mechanisms.

Cholestyramine and colestipol - effectively lower serum TC, especially when coupled with diet. A dosage of 12 to 32 gm orally in 2 to 4 divided daily doses will lower LDL levels by 25 to 50%. Side effects - constipation.

Niacin (nicotinic acid) may also be useful in HLP, but the high dosage required (3 to 9 gm/day orally in divided doses with meals) coupled with its side effects (gastric irritability, hyperuricemia, hyperglycemia, flushing, and pruritus) often restricts its use. Niacin is most effective when combined with cholestyramine.

Lovastatin (an HMG-CoA reductase inhibitor) - 20 - 80 mg orally in 1-2 daily doses can remarkably lower LDL level. Its effectiveness can be enhanced when combined with cholestyramine and/or niacin. Side effects - hepatitis and myositis (more often if combined with gemfibrozil, clofibrate, or niacin).

Probucol - 500 mg orally bid lower LDL levels 10 to 15% when added to diet, but it often has the additional undesirable side effect of lowering HDL levels.

Thyroid analogs (D-thyroxine) - effectively lower LDL levels but are contraindicated in patients with suspected or proven heart disease.

Clofibrate - has little effect on plasma TC or LDL levels in this disorder, may produce gallstones and metabolic problems.

Treatment of hypertension emergency state (crisis)

Sodium nitroprusside – given in continuous IV infusion (not more than 10 min in higher dose). Producing venodilatation and atreriodilatation.

Side effects – nausea, vomiting, agitation, muscle twitching, acute psychosis (because of thiocyanate intoxication – antidote have to be given – thiosulfate Na)

 **OPIOID DRUGS**

Definitions

1.	All drugs, including opiates such as morphine and codeine, which have morphine-like activity

2.	Often mistakenly called narcotic drugs, which now refers mainly to abused substances

3.	Endogenous opioids are divided into three groups: enkephalins, endorphins and dynorphins.

Mode of action

1.	There are three classes of opioid receptors involved in analgesia – m, d and k.

a.	m receptors are the most common sites of action; respiratory depression, euphoria, miosis and constipation appear to be mediated by these receptors. The endogenous substrates are b-endorphin and the enkephalins. Selective agonists are morphine, methadone and fentanyl, among others.

b.   The endogenous substrate for d receptors is the enkephalins.

c.   The endogenous substrate for k1 receptors is dynorphin A; for k3 receptors it is not known.   Dysphoria and disorientation or depersonalization are mediated through these receptors.

2.   Nonselective antagonists for the opioids are naloxone and naltrexone. They affect all three types of receptor.

Uses

1.   Mainly as analgesics; since there is potential for abuse, tolerance and dependence, opioids should be reserved for use in severe pain or in the treatment of patients with terminal illness. For chronic pain not due to malignancy, it is best to use other forms of analgesia. For patients with terminal cancer, regular small doses throughout the day have been shown to be more effective than administration only on demand. Morphine is the drug-of-choice in this situation and oral preparations are very effective.

2.   Morphine is also used in the treatment of pulmonary edema or acute left ventricular failure.   Dyspnea is relieved, probably due to vasodilatation and decreased peripheral resistance.

3.   Codeine and dextromethophan are used as cough suppressants.

4.   Meperidine derivatives are used in the symptomatic treatment of diarrhea.

5.   Fentanyl, sufentanil and alfentanil are used in general anesthesia.

6.   Naltrexone has been found useful in the treatment of alcoholism.

Routes of administration, absorption and metabolism

1.   Oral – including immediate release and sustained release; good absorption

2.   Other methods include: intravenous, intramuscular, intrathecal, epidural, subcutaneous, sublingual and rectal suppositories.

3.   Codeine, heroin and methadone cross the blood-brain barrier more easily than morphine.

4.   Metabolism is in the liver.

General side-effects

1.   All opioids have a tendency for the development of tolerance and physical dependence.   Gradual decrements in dose will prevent withdrawal.   Tolerance is both to the beneficial effects and to the side effects.

2.   Hypersensitivity is uncommon and anaphylactoid reactions are rare.

3.   Respiratory depression causes CO2 to accumulate, which causes increased blood flow in the brain and increases intracranial pressure.

Toxic reactions

1.   Triad of coma, pinpoint pupils and respiratory depression is typical.

2.   Reduced state of consciousness - stupor or coma, bradypnea, cyanosis

3.   Blood pressure soon falls. As a result, urine production is decreased. There is also a decrease in body temperature and skin becomes cold and clammy.

4.   Severe hypoxia will cause dilatation of pupils.

5.   Seizures are often seen in infants and children.

6.   Muscles are flaccid and the tongue may block the airway.

7.   The first step of treatment is to establish an airway and ventilation.

8.   Specific treatment is by naloxone – give slowly at first to avoid withdrawal, keeping a close eye on the patient until respiratory function returns to normal.

Contraindications

Relative

1.   Cor pulmonale (already compensating for decreased respiratory function)

2.   Liver failure

3.   Head injury (elevated intracranial pressure already exists; mental clouding with miosis may cause diagnosis to be missed)

4.   Use of morphine, codeine, propoxyphene or meperidine (pethidine) in renal insufficiency

5.   Best not to use morphine or codeine during acute asthmatic attack due to suppression of cough reflex.

Absolute

Do not give meperidine (pethidine) with monoamine oxidase inhibitors – can cause profound respiratory depression, respiratory excitation, convulsions, high fever and/or delirium.

Drug interactions

1.   Opioid effects may be increased by phenothiazines, monoamine oxidase inhibitors and tricyclic antidepressants.

2.   Rifampin and phenytoin cause decreases in methadone concentrations.

3.   Chlorpromazine and tricyclic antidepressants intensify the respiratory effects of meperidine (pethidine). Chlorpromazine and promethazine intensify the sedative effects of meperidine (pethidine).

4.   The addition of amphetamines enhances the analgesic effects of opioids and decreases the sedative effects. The addition of antidepressants, such as amitriptyline or desipramine, enhances the analgesia, also.

Examples

I.   Opioid agonists

Morphine

1.   Morphine is the prototype for opioid agonists.

2.   Work mainly via m receptors; high doses may also cause effects via other receptors

3.   More effective against dull, constant pain than against acute, colicky pain; mostly used for severe pain or in terminal patients

4.   Effects include

a.   analgesia without loss of consciousness, drowsiness, alterations of mood and cloudy mental processes, euphoria

b.   miosis – pinpoint pupils is a pathognomonic sign

c.   respiratory depression, inhibition of cough reflex

d.   decreased peristalsis, constipation, nausea and vomiting

e.   peripheral vasodilatation causing decrease in peripheral resistance and orthostatic hypotension, flushing

Codeine

1.   Effective orally

2.   It converts to morphine, which is the likely reason for its analgesic effects.

3.   Used as a cough suppressant and analgesic; often found in preparations with acetaminophen (paracetamol) or aspirin

Methadone

1. Mostly used to treat heroin addicts by replacing the heroin with a satisfactory (to the addicts) substitute which is less addictive and less dangerous; less common uses are in analgesia and treatment of withdrawal syndromes.

2. Oral absorption is good.

3. Side effects and toxicity as for morphine. With long-term use, there may be hyperhidrosis, lymphocytosis, prolactinemia and high levels of protein in the blood.

Propoxyphene

1. Analgesic and CNS effects as morphine; GIT effects as codeine

2. Oral absorption is good; intravenous and subcutaneous administration are very irritating

3. Used for analgesia and most often given in preparation with aspirin or acetaminophen (paracetamol)

4. Toxic reactions include seizures, respiratory depression, delusions, hallucinations, cardiotoxicity and pulmonary edema.

Meperidine (pethidine)

1.   Used for analgesia, especially in childbirth

2.   Crosses the placenta; if given to the mother too close to delivery, neonate may need naloxone.

3.   Effects are sedation, respiratory depression and euphoria as with morphine. Intravenous administration may cause tachycardia. Less constipation and urinary retention than with morphine.

4.   Toxic reactions include tremors, seizures, hyperreflexia and muscle twitching.

Loperamide and diphenoxylate

1.   derivatives of meperidine

2.   Major effect is on slowing of GIT passage in the treatment of diarrhea

3.   Given orally; low systemic effects as most is excreted in the feces (and thus high concentrations get to the target organ)

Fentanyl

1.   Used in general anesthesia, often in combination with the neuroleptic droperidol, and for postoperative analgesia

2.   Administration is intramuscular or intravenous, but also can be via transdermal patches (for pain only).

3.   Derivatives such as sufentanil and alfentanil can be also used via intrathecal or epidural routes.

4.   Toxic effects include muscular rigidity, which responds to naloxone.

Others include papavarine (used as a smooth muscle relaxant and for visceral pain), apomorphine (an emetic and dopamine agonist) and heroin.

II. Opioid antagonists

1.   Have minimal effect unless opioids have been administered, or the body's endogenous opiates activated

2.   Used mainly to treat overdoses and in neonates exposed in utero to opioids

3.   Naloxone and naltrexone are not specific; naloxone is the most commonly used. Levallorphan and nalorphine are selective antagonists at m receptors, but agonists for k receptors (and so have analgesic properties, as well).

4.   Naloxone is not given orally, as the first pass effect metabolizes it all. Thus it must be given parenterally only. Naltrexone can be given orally.

5.   Naloxone may cause rebound effects – i.e. tachypnea and catecholamine release. The sudden elevation in catecholamine levels may cause arrhythmias.

6.   Naloxone may cause withdrawal effects in chronic users of opioids.

III. Mixed agonist and antagonist effects

1.   Effects depend on which receptor is activated

2.   Respiratory depression, but not psychological effects, is inhibited by naloxone.

3.   Nalbuphine is a competitive antagonist at the m receptor, but acts as an agonist at the k receptor, thereby causing analgesia, its main effect. Cyclazocine and nalorphine are similar.

a.   Causes analgesia and respiratory depression as with morphine.

b.   It can cause withdrawal in chronic users of opioids.

c.   Side effects include GIT symptoms, sedation, sweating and headache.

4.   Pentazocine is a weaker m antagonist and is less likely to result in abuse than stronger agonists such as morphine, especially when given orally. Similar is butorphanol.

a.   Well absorbed orally and parenterally; crosses the placenta

b.   Its effects are analgesia, sedation and respiratory depression.   It is used as an analgesic for severe chronic pain

c.   Side effects are as morphine on the GIT, but pentazocine causes elevated pulse and blood pressure.   Most commonly seen are nausea, sedation, sweating and dizziness.   May cause withdrawal symptoms in chronic users of opioids.

d.   Toxic effects include hallucinations, nightmares, anxiety, respiratory depression (can be treated by naloxone), tachycardia and hypertension.

5.   Buprenorphine is a partial m agonist. Its effects and side effects are as with morphine.

18 NSAIDs - NONSTEROIDAL ANTIINFLAMMATORY DRUGS

Definition

Drugs that are not steroids, but have similar anti-inflammatory effects; also are antipyretic and analgesic

What is a Steroid?

Steroids are lipids derived from cholesterol. Testosterone is the male sex hormone. Estradiol, similar in structure to testosterone, is responsible for many female sex characteristics. Steroid hormones are secreted by the gonads, adrenal cortex, and placenta.

Structure of some steroid hormones and their pathways of formation. Images from Purves et al., Life: The Science of Biology,

Mode of action of NSAID

1.   Inhibition of cyclooxygenase, and so inhibition of production of prostaglandins and thromboxane from arachidonic acid (but not leukotrienes, which are produced by another enzyme called

lipooxygenase). Prostaglandins are essential for inflammatory reactions and fever. NSAIDs effect on pain is mainly on pain caused by inflammation and other situations where prostaglandins are elevated.

2.   Aspirin causes irreversible inhibition by acetylating the cyclooxygenase.

3.   Others compete with arachidonic acid at the active site on the cyclooxygenase.

Uses

1.   inflammatory conditions such as rheumatoid arthritis, osteoarthritis, gout, ankylosing spondylitis

2.   pain relief – from mild to severe; not good for visceral pain, but especially helpful in dysmenorrhea and postoperative pain

3.   reduction of fever by returning the hypothalamus set point to its normal value

4.   closure of patent ductus arteriosus in neonates (especially indomethacin)

5.   inhibition of uterine motility – i.e. delay uterine contractions in premature labor, treat dysmenorrhea

Routes of administration, absorption and metabolism

1.   Oral, rectal (ketorolac can also be given intramuscularly)

2.   Good oral absorption

3.   Excreted by the kidneys

General side effects

1.   Gastrointestinal – ulcerations, bleeding, heartburn

2.   Hypersensitivity – mostly to aspirin – manifesting as allergic rhinitis, angioneurotic edema, urticaria or asthma, usually in middle-aged people with asthma, nasal polyps or chronic urticaria

3.   Disturbances of renal function in patients with congestive heart failure, cirrhosis, hypovolemia or chronic renal disorders

4.   Chronic excess use can cause chronic interstitial nephritis and papillary necrosis

5.   Inhibition of platelet aggregation (due to lack of thromboxane) – can cause increased bleeding time

Contraindications

1.   Do not give any NSAID to patient with allergic reaction to aspirin; a reaction similar to anaphylactic shock could result.

2.   Do not give aspirin to child with influenza or chickenpox – risk of Reye's syndrome.

3.   Do not give indomethacin to patient with renal or peptic ulcer disease, epileptics, patients with Parkinsonism or psychiatric disorders, nursing mothers or pregnant women (unless needed for premature contractions).

4.   Do not give phenylbutazone to patients with hypertension, peptic ulcer disease, or with cardiac, renal or hepatic disorders.

5.   Don't give aspirin to patient with chronic hepatic disease, vitamin K deficiency, hemophilia or hypoprothrombinemia – platelet problems may cause hemorrhage.

Drug interactions

1.   Use of salicylates or phenylbutazone at the same time as warfarin, sulfonylureas or methotrexate should be avoided. If necessary to give both, dosages will need to be adjusted since the NSAIDs compete with these drugs for binding proteins.

2.   Indomethacin concentration is increased with administration of probenecid, but dose does not need to be adjusted.

3.   Indomethacin blocks the therapeutic effects of furosemide on hypertension and sodium.

4.   Piroxicam reduces the clearance of lithium from the kidneys.

Examples

I. Salicylic acid derivatives – aspirin, diflunisal, salicylsalicylic acid, sulfasalazine, mesalamine

1.   Aspirin still most widely-used

2.   Oral is the best form of administration; rectal absorption is not reliable. Some topical preparations are available. Metabolism is in the liver.

3.   Uses

a.   Antipyretic, analgesic and antiinflammatory; especially good for pain of headaches, muscles and joints

b.   Aspirin also has antiplatelet activity since it causes irreversible inhibition of both cyclooxygenases and platelets are not able to synthesize cyclooxygenase. Thus, until the next generation of platelets develops, which takes 8-11 days, platelet function is disturbed. This enables aspirin to be used for prophylactic treatment of coronary artery disease, thrombotic CVA and hypercoaguable states.

c.   Topical use of salicylic acid irritates skin and mucus membranes and so is used in the treatment of warts, corns and some fungi.

d.   Rectal suspensions and suppositories of mesalamine are useful in the treatment of inflammatory bowel disease. Sulfasalazine is ingested and in the colon breaks down to mesalamine.

4.   Side effects

a.   Side effects include heartburn, upper gastrointestinal bleeding, nausea and vomiting, hyperventilation with respiratory alkalosis; aspirin may cause mild hemolysis in patients with G6PD deficiency. GI bleeding may be painless and so discovered late.

b.   In children, metabolic acidosis mixed with respiratory acidosis may follow the respiratory alkalosis, causing hypernatremia, hypokalemia and dehydration.

c.   Even in healthy patient, moderate use of aspirin doubles bleeding time and so aspirin should be discontinued at least a week before any surgery.

5. Toxic reactions

a.   Toxic doses can cause convulsions, tinnitus, deafness, psychosis, stupor and coma – which are reversed with withdrawal of the drug. Also seen are respiratory depression with respiratory acidosis, aminoaciduria, hyperglycemia and glycosuria. Fever and accompanying dehydration is common, especially in children. Children may also show hypoglycemia

b.   Salicylate intoxication is an emergency (which can be fatal) and patient needs to receive fluids, ventilation and cardiovascular support. Activated charcoal will neutralize aspirin still in the GIT. Vitamin K, bicarbonate and plasma may be given as needed. In a few cases, hemodialysis may be necessary to rid the body of the salicylate.

II. Indole and indene acetic acids

Indomethacin

1.   Has similar therapeutic characteristics to the salicylates, but also inhibits neutrophil mobility.

2.   Good oral absorption; good concentration in synovial fluid makes it a good drug for joint problems.

3.   Up to half of patients who take indomethacin complain of side effects and about 20% stop the drug because of this. Most effects are dose-related. Because of these adverse effects, indomethacin is not used widely.

4.   It uses include cases of refractory fever or inflammation which do not respond to other drugs. It also has uses in the treatment of premature uterine contractions and in closing the ductus arteriosus in neonates. It has also been used in Bartter's syndrome.

5.   Adverse effects include

a.   GIT – nausea, anorexia, abdominal pain, ulcerations, GI bleeding, acute pancreatitis, diarrhea

b.   Severe headache in the frontal area of the head is seen in 25-50% of patients who take indomethacin chronically. Other CNS symptoms include dizziness, vertigo, confusion and a feeling of lightheadedness.

c.   Hematological side effects include neutropenia and thrombocytopenia. Aplastic anemia is rare.

d.   Hypersensitivity manifests as rash, urticaria and asthma.

Sulindac

1.   Similar to indomethacin, but has less undesirable effects and also is much less potent

2.   Sulindac is a pro-drug – the active metabolite is sulfide

3.   Used in inflammatory conditions; may also be used in premature uterine contractions

4.   Side effects include nausea, abdominal pain, dizziness, sleepiness, headache, nervousness, rashes and pruritus

Etodolac

1.   new on the market

2.   Good oral absorption

3.   Used in inflammatory conditions and after surgery

4.   Minimal side effects – GIT ulceration, rash, CNS effects

III. Heteroaryl acetic acids

Tolmetin

1.   Good oral absorption, excreted in the urine

2.   Used in inflammatory conditions

3.   Side effects occur in up to 40% - usually epigastric pain, dyspepsia, nausea, vomiting, ulceration. CNS effects are less than with indomethacin or aspirin

Ketorolac

1.   Can be given parenterally – both oral and intramuscular absorption are rapid and good

2.   Analgesic activity much more than antiinflammatory; used as antiinflammatory agent in the eye and for pain after surgery

3.   Side effects include sleepiness, dizziness, headache, dyspepsia, nausea, abdominal pain and pain at the injection site

Diclofenac

1.   Oral absorption is rapid and complete, but 50% is lost to first pass effect; metabolism is in the liver

2.   Used long-term in inflammatory conditions, but also useful in acute painful conditions including dysmenorrhea, bursitis and after surgery. An ophthalmic formulation is used after cataract surgery. A compound containing diclofenac and misoprostol (PGE1 analog – to reduce GIT side effects) is on the market and is effective with reduced side effects.

3.   Side effects appear in 20% of users and are mainly GIT. Also seen are CNS effects, rashes, edema and transient elevations in liver enzymes.

IV. Arylpropionic acids

In general

1.   in wide use, are better tolerated than aspirin and indomethacin

2.   used for inflammatory conditions and as analgesics (musculoskeletal pain, dysmenorrhea)

3.   like aspirin, affect platelet function

Ibuprofen

1.   Rapid oral absorption

2.   Side effects are mostly GIT (but less than aspirin or indomethacin).   Less common are thrombocytopenia, headache, blurred vision and skin rash. Rare are edema and toxic amblyopia.

Naproxen

1.   Complete oral absorption; rectal absorption is slower

2.   Side effects are GIT and CNS as in other NSAIDs

Others include flurbiprofen, ketoprofen, fenoprofen and oxaprozin

V. Fenamates – mefenamic acid, meclofenamic acid

1.   Not widely used since no advantages over other NSAIDs and significant side effects

2.   Good oral absorption

3.   Side effects are mostly of the GIT – dyspepsia, diarrhea, steatorrhea – which are seen in about ¼ of patients taking these drugs.

4.   Hemolytic anemia has been seen

VI.   Enolic acids

Piroxicam

1.   Long half-life – can be given once a day; complete and fast oral absorption

2.   Also inhibit activation of neutrophils

3.   Used mostly in rheumatoid arthritis and osteoarthritis

Phenylbutazone

1.   Severe side effects have limited its use to antiinflammatory indications and only after other drugs have failed.

2.   Rapid and complete oral absorption

3.   Up to 45% suffer from side effects – most common are GIT effects; also seen are hypernatremia, hyperchloremia and edema with increased risk for pulmonary edema

4.   More serious effects include hypersensitivity similar to serum sickness, hepatitis, nephritis, aplastic anemia, agranulocytosis, leukopenia and thrombocytopenia

Nabumetone

1. Fewer side effects than most NSAIDs, especially less dangerous to the stomach; due to its being a prodrug and also to selective inhibition of cyclooxygenase-2 which has less effect on the GIT.

2. Used in inflammatory conditions

3. Side effects include rash, headache, heartburn, dizziness, tinnitus and pruritus.

Choice of drug in special situations

1. In children, use naproxen or tolmetin. Aspirin has been associated with Reye's syndrome.

2. In pregnant women, avoid NSAIDs, especially in the late third trimester. If necessary, small amounts of aspirin may be given.

PARACETAMOL (ACETAMINOPHEN)

Although not actually an NSAID, paracetamol (acetaminophen) has similar properties to the NSAIDs and so is covered here.

Uses

Weak antiinflammatory effects – used as antipyretic and analgesic

Routes of administration, absorption and metabolism

1.   oral or rectal

2.   rapid and almost complete absorption

Side effects

1.   Unlike aspirin, no effect on GIT mucosa or platelets

2.   May see urticaria, but in general minimal side effects

3.   Hepatotoxicity, however, is seen with overdose (intentional or otherwise).

a.   It begins with nausea, vomiting, anorexia and abdominal pain. Within 2-4 days there are elevated liver enzymes, high bilirubin and increased prothrombin time.

b.   If not treated, hepatic failure may occur and also renal failure. The likelihood of hepatic damage depends on the blood concentration and how long since the dose was taken. A graph compares these two parameters and is used to determine prognosis.

c.   Treatment consists of gastric lavage – within 4 hours to be effective, supportive treatment and the specific antidote – N-acetylcysteine (can be given up to 36 hours after ingestion, but is most effective in the first ten hours)

www.ingramcontent.com/pod-product-compliance
Lightning Source LLC
Chambersburg PA
CBHW031828170526
45157CB00001B/225